Quantum Physics for Beginners

K.I.S.S. 'n Tell

A Keep It Simple Short Tale To Understand The Secrets And The Fundamental Laws Of The Universe Through Its Compelling Story. Almost No Math Involved!

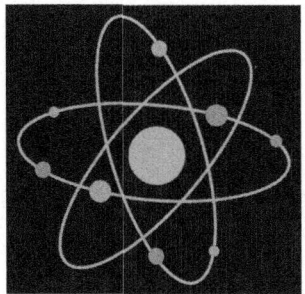

Antonio Scalisi – Karing Ship

Special Thanks to Silvia

© Copyright 2022 Antonio Scalisi - Karing Ship
All rights reserved.

The content contained within this book may not be reproduced, duplicated or transmitted without direct written permission from the authors or the publisher.
Under no circumstances will any blame or legal responsibility be held against the publisher, or authors, for any damages, reparation, or monetary loss due to the information contained within this book. Either directly or indirectly.

Legal Notice:
This book is copyright protected. This book is only for personal use. You cannot amend, distribute, sell, use, quote or paraphrase any part, or the content within this book, without the consent of the authors or publisher. It is possible to share only the images indicated in the Final Notes (in accordance with the terms of the specific licenses)

Disclaimer Notice:
Please note the informations contained within this document are for educational and entertainment purposes only. All effort has been executed to present accurate, up to date, and reliable, complete information. No warranties of any kind are declared or implied. Readers acknowledge that the author is not engaging in the rendering of legal, financial, medical or professional advice. The content within this book represents the personal vision of the authors on the subject dealt with. Please consult a licensed professional before attempting any techniques outlined in this book.
By reading this document, the reader agrees that under no circumstances are the authors or the publisher responsible for any losses, direct or indirect, which are incurred as a result of the use of information contained within this document, including, but not limited to, — errors, omissions, or inaccuracies.

Table of Contents

Introduction..8

Chapter 1. Wave It Is...12

Chapter 2. XIX Century, Classical Physics' Last Two Big Hits..20

   2.1 – An Endless Stream of Achievements....................20
   2.2 - Thomas Young's Experiment - 1800......................21
   2.3 – Electromagnetic Waves...27
   2.4 – The Speed of Light..30

Chapter 3. End of the Century – Mother Nature Strikes Back..34

   3.1 – Photoelectric Effect: the Problem.........................34
   3.2 – Hertz Experiment in your Kitchen........................36
   3.3 – Planck and the Ultraviolet Catastrophe...............40
   3.4 – Photoelectric Effect: the Solution........................43

Chapter 4. De Broglie (1924) – The Wave in You..............46

   4.1 - De Broglie's Ph.D Thesis.......................................46
   4.2 - 1927 Davisson & Germer's Experiment.................47

Chapter 5. Born, Heisenberg, Schrödinger & His Cat.......54

   5.1 - Max Born..54
   5.2 – The Double Slit Once Again (1961).....................55

5.3 - Copenhagen Interpretation (1927)..........................60
5.4 - Heisenberg & Schrödinger's Cat...........................62

Chapter 6. Atoms – Quantum Mechanics' Cup of Tea......68

6.1 – Atom's Lifetime...68
6.2 – Bohr's Atomic Model..69
6.3 - Spectroscopy...71
6.4 – From Orbit to Orbitals..74
6.5 – So What?..79

Chapter 7. Practical Applications.........................82

7.1 – Electronic Devices...83
7.2 – The Thermionic Valve..85
7.3 – Off the Record – Two Electronic Circuits............87
7.4 - Semiconductors and Diodes...............................89
7.5 – Quantum Entanglement and Friends..................94
7.6 – Quantum Cryptography.....................................95

Chapter 8. From Waves to Strings..........................98

Chapter 9. True or False? Many Universe Theory. The Law of Attraction. God's Equation...........................104

9.1 Multiverse or Many Universe Theory...................105
9.2 The Law of Attraction...107
9.3 God's Equation & "Partners"..............................110
9.4 The Theory of Everything...................................117

Conclusions..120

Notes..124

Image Credits..124
Text Notes...124

About the Authors..126

Antonio Scalisi..126
Karing Ship...127

Other books by the Publisher..128

Introduction

"If you think you understand quantum mechanics, you don't understand quantum mechanics".

Folks, I'm not trying to insult you, but this statement is attributed to Richard Feynman, the father of Quantum Electrodynamics, the theory for which he won the Nobel Prize. This is a theory that applies quantum mechanics to all phenomena in which the forces involved are the electric and / or magnetic forces. These are – just between you and me - 99.999% of those we observe every day, like for example: the forces that explain the fact that at this moment you are not sinking through the chair you are sitting on, the beating of our heart, how mother Nature arranges atoms in molecules, how light behaves when it interacts with bodies and so on and so forth.
So why should you read this book?
Because the meaning of the above statement can be better outlined by another Feynman's famous statement:

"Quantum mechanics describes nature as absurd from the point of view of common sense. And yet it fully agrees with experiments. So I hope you can accept nature as She is - absurd."
For those who care, there are many other such statements by those who contributed to the development of this theory; just type "quantum mechanics quote" on your preferred search engine.
Let's go back to the first question: why reading this book, or any of the books that "claim" to be willing to explain quantum mechanics, and moreover, to the layman?
Well it is obvious!
Our goal is to become fully aware that "the only true wisdom is in understanding we understand nothing ", just as Socrates came to say "*the only true wisdom is in knowing you know nothing*".

In this book you will go through all (well let's say the main ones) troubles that forced many great scientists to free themselves from the constraints of logic and common sense; together with them you will build this fascinating picture of how, for now, it is believed that Mother Nature really works.

No math involved!

We will therefore follow a historical approach, starting from the most amazing successes of classical physics (19th century): Maxwell's equations, the discovery of electromagnetic waves and the affirmation of the wave theory of light. Then, starting from the second half of the same century, we will learn about the first failures: the photoelectric effect and the ultraviolet catastrophe, which led (in 1905) a then unknown Albert Einstein to write an article (he wrote 4 but we are only interested in one of those) where he claimed a dual nature of light, both wave and particle nature; this is the article that gained him his Nobel Prize. After Einstein, we will meet in 1928 a French doctoral student named Louis De Broglie, who advanced the hypothesis that ALSO the constituents of matter, - the only ones known at that time were the electron and the proton-, must possess this wave-particle dualism. Then in the company of the two American physicists from a few years later, Clinton Davisson and Lester Germer, you will perform the first experiment that will confirm this hypothesis. From there, the story will lead us to interpret some properties of atoms - including their very existence – thanks to Niels Bohr, a Danish physicist, who actually published his ideas a few years before this wave-particle dualism. However, his own theory ended up fitting it perfectly. Erwin Schrödinger will help us putting everything in order so that we can then launch ourselves towards the subsequent triumphs of the theory (and its practical achievements).

An essential requirement to be able to follow this path without getting lost is to be convinced of what is implicitly expressed in the second of the Feynman's quotes I reported: you must immediately convince yourself that science, and therefore also physics, does not explain the **WHY** of things, but it is concerned with describing **HOW** Mother Nature works. A scientific theory never explains why, but it rather produces quantitative predictions (what speed a certain body reaches when it falls from x yards, what I have to do to get a certain amount of energy from a pound of uranium, etc.). Science therefore doesn't even need to be Logical. The success or failure of a theory lies in the precision and simplicity with which it predicts the greatest amounts of phenomena, and to date, quantum mechanics (with all its derivations) is the theory that, despite its absurdities, has beat all the others.

We finally reached the end of this introduction, and as Dante (you) with Virgil (me) we begin this fascinating journey through time and physics.

Chapter 1. Wave It Is

Let's start with the assassin: quantum mechanics will claim that an electron (or a proton or ... even yourself) is a wave, just like a photon is, - perhaps this one you will have more or less already heard of.

So, in order to be able to well and clearly capture what lies behind what I have just written, we need to spend some time to see (for many of you as a repeat) some properties of waves. Any wave.

In truth, we will focus mainly on that particular property of waves which is called interference, but let's proceed in order.

Let's define what a wave is; for this purpose, we resort to a practical example.

If I take a string at one end and start moving my hand up and down with some regularity, the string will take on a shape like the one in the figure 1.1 below, which represents the string in three different moments in time.

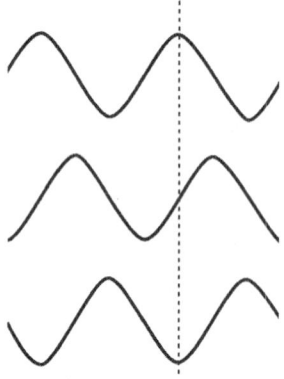

Fig.1. 1

I took the liberty of choosing a random point, where I drew a (vertical) dotted line, to highlight that, at different instants, a certain property of that point (in this case the string vertical position) varies, periodically oscillating from "all the way up" to "all the way down".

I used the generic expression "a certain property of that point" because there are other types of waves where we don't have a piece of string going up and down, but, for example, we have air pressure oscillating between a maximum and a minimum. You all know this second type of wave: we call it SOUND and from a mathematical viewpoint it is identical to the case of the string, even if the "property" that oscillates is no longer the vertical position of a piece of string, but the air pressure at any particular point in space.

What I want to stress here, is the concept of what "wave" means: a set of values, of a certain property of points of space, where not all points may take on the same value at the same time and where the values fluctuate over time with a certain regularity, between a maximum and a minimum.

As the string is easier to visualize, let's go back to this example even if, I REPEAT IT, what we are about to say applies to ANY KIND OF WAVE!

First, just for completeness, a wave travels in time and space at a certain speed. If the string were 50 km long, it would take some time before someone sitting at the other end would begin to see it swinging. Each wave has its own speed, which is called "wave propagation speed". After clarifying this, we (almost) won't talk about it anymore.

Instead, there is another property that is much more important to us, which is called "wavelength": what is it?

Let's review the previous figure, which I report below (fig. 1.2), where I removed the vertical hatching, but I highlighted with double arrows the distance between two maximums or between two minimums or between two "similar" points.

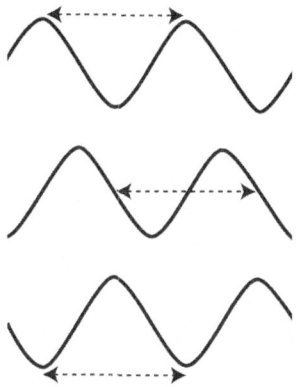

Fig 1.2

This distance, which as you can see is identical for the three pairs of points, is called "wavelength". Each wave has its own wavelength, that represents the distance between the two closest points in space that at a certain moment share the same property: in this case the two closest maxima or the two closest minima or the two most "similar" neighbors.

But, you could tell me: "Listen pal, the two closest "similar ones" are not the ones you indicate, there is one even closer". And I tell you that it could look like you are right, but it is like I say! Because what you believe to be "similar" and closer has the same vertical position but, unlike the ones I chose, it is pointing uphill while the ones I choose are both pointing downhill!

So, I hope I have made it clear what this wavelength is.

There is a third property that I must mention because sometimes we will use it in place of the wavelength, it is called "frequency" and it is not independent from the wavelength. Let's go back to the first figure that I re-propose below (fig. 1.3): I hope you notice that the wave is moving to the right hand side.

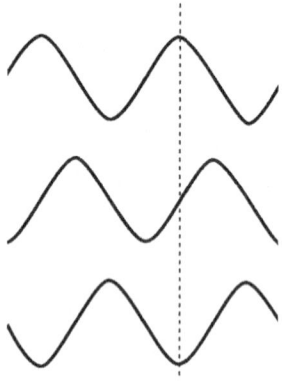

Fig. 1.3

So, frequency is the measure of how many times over a certain period, let's make it a second (or a minute, an hour, as it pleases us), any point in space, for example where I once again drew the vertical dotted line, is crossed by a maximum or a minimum or a "similar". Now I hope you know what I mean with this "similar".

I hope it is clear that, given a certain wavelength, the higher the wave propagation speed (here it comes) the higher the number of maxima (or minima, etc.) that cross this point over a period of time. Indeed, the faster the wave moves, the faster the second maximum will reach the dotted line, followed by the third one and so on. Or else, given a certain speed, the shorter the wavelength, AGAIN, the higher the number of maxima that cross the dotted line in a second or so.

Mathematically, if I call "**v**" the propagation speed and "**λ**" the wavelength, it turns out that the relationship that binds them to the frequency, which for convenience we indicate here with the letter "***f***", is the following: ***f*****=v/λ**. Looking at this formula it should be clear that the larger λ, the smaller the frequency, or else, the greater the speed v and the greater the frequency. In short, the concept you should keep in mind is that wavelength and frequency are bound together, and as one increases the other decreases. So why bother introducing it!? Well, we will see that sometimes it is more convenient to use one and sometimes to use the other.

Now let's get to the guts of this chapter, let's take two waves: one coming from the left hand side and one from the right hand side, as in the following figure 1.4

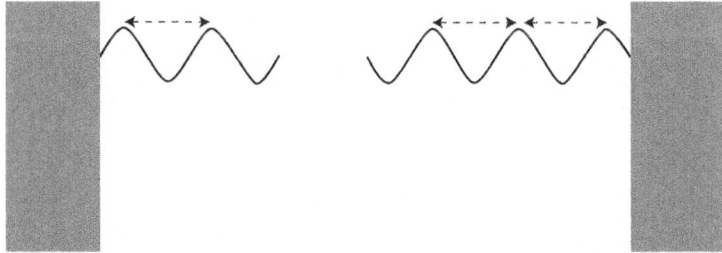

Fig. 1.4

The two waves are identical: they have the same wavelength, the same propagation speed and are also in phase. What does "in phase" mean? If you look at how their origins behave, they happen to do the exact same thing at the same time: when one goes up the other goes up and likewise when they go down.

Now let's take a random point P on the left hand side wave, highlighted by the vertical hatching in the figure 1.5 below.

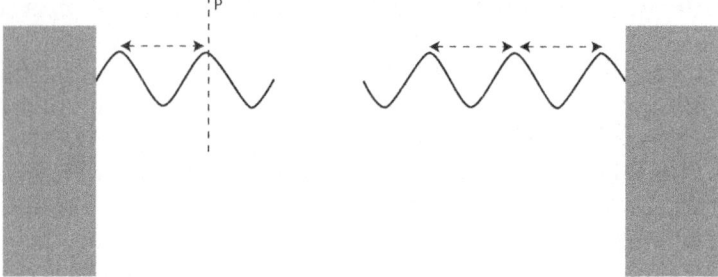

Fig. 1.5

Let's assume that the wave from the right hand side, instead of originating at a distance from P, like in the case of the above figure, originates from a particular distance as in the figure 1.6.

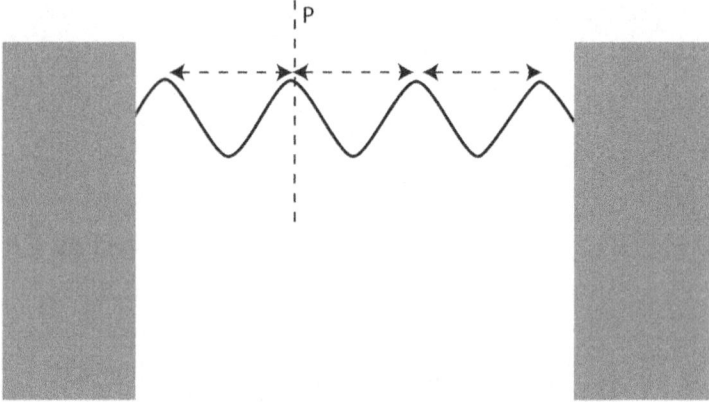

Fig. 1.6

I hope it is clear that point P receives an identical contribution from both waves, therefore its oscillation will be equal to the sum of the two oscillations.

What is so particular about this drawing? Well, the difference between the distances of point P from the right and the left sources is exactly equal to one wavelength.

Moreover, I hope it is clear that the same result would have been obtained if the difference between the distances were exactly two, three, or four wavelengths: we just need it to be an INTEGER number.

Now let's change the scenario and move the source on the right hand side a little far to the right, as in the figure 1.7 where for more clarity (I hope) I changed the color of the right hand side wave, which is now red*

*Note:(*only in some versions of eBook readers).*

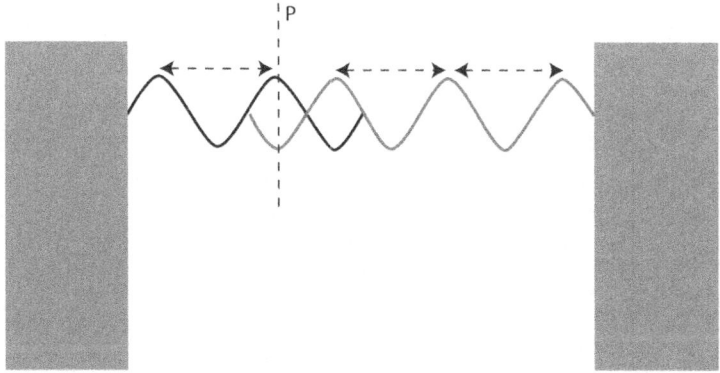

Fig. 1.7

I hope it is clear that this time, when the left wave pushes "up" point P, the right one pulls it "down" and so on. The two waves will always do the opposite of each other with the result that P will NEVER undergo any oscillation at all!
But what did we change from the previous design? The difference between the two distances has changed! We now have an additional half wavelength. In other words, the difference between the two distances of each source from P is half a wavelength, or a length and a half, or two and a half, etc.
Let's wrap it up. If the difference between the distances the two waves had to travel to reach point P is an integer multiple of the wavelength, we have the sum of the contributions, what in jargon is called "constructive interference"; if, on the other hand, a half wavelength is also added, the contributions are neutralized, and we call this "destructive interference".
Please be advised, this obviously can happen to any point in space (not just in one dimension like in the example) when we have IN PHASE waves starting from two (or even more than two) different sources and propagating in any direction, as in the arrows in the figure 1.8 below.

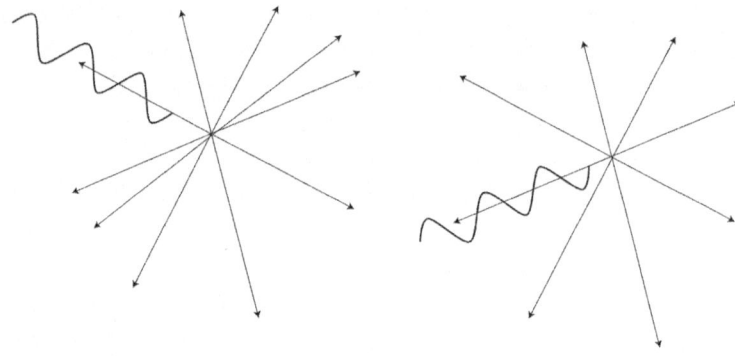

Fig. 1.8

So what?

Well, I would like to convince you that, FOR ANY KIND OF WAVE, when there are two or more identical "in phase" waves there will ALWAYS be a set of VERY PRECISE spots in space where you will ALWAYS have maximum oscillation, and others where there will ALWAYS be zero oscillation. Furthermore, the location of these spots depends exclusively on the WAVELENGTH and on the positions of the spots in relation to the two sources, since what matters is only the difference, expressed in wavelengths, of the paths that each wave must travel to reach that particular spot from its own source. All this should be clearly visible in the picture 1.9 below, which shows two identical and in phase water waves.

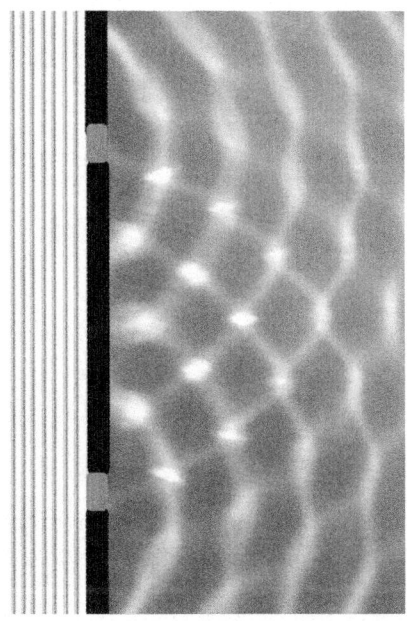

Fig. 1.9

As you can see, there are areas made up of maxima only, which alternate "radially" with those made up of minima only.
We are done with wave theory!
Let's just try to keep in mind how we came to understand that both constructive and destructive interference exist, because these concepts will often return in following chapters. However, bear in mind that there are scenarios where we can get the same interference effect, with a similar math but a little more complicated, therefore I won't cover them.
And now we can happily dive into the history of quantum mechanics.

Chapter 2. XIX Century, Classical Physics' Last Two Big Hits

2.1 – An Endless Stream of Achievements

Classical physics, or shall we say Modern Science, was born in the course of the 17th century thanks to the genius of Galileo Galilei.
In actual fact, before Galileo, science - better yet, pseudo-science- was influenced by Aristotle's approach, not exactly an idiot. His approach had indeed never been questioned for a good two thousand years.
However, Aristotle was not a scientist in the modern sense of the term, precisely because his goal was not to create models to provide quantitative predictions of HOW Mother Nature appears to behave. Aristotle wanted to explain the WHY of the phenomena we all observe. Further, he wanted the explanations to be LOGICAL!
Let's take for example one of Aristotle's biggest blunders and see the genius with which Galileo (I repeat, after two thousand years) overcame it.
Like all of us, Aristotle had noticed that heavier bodies fall to the ground sooner than lighter ones. But, far from trying to create a model that could predict how quickly a certain body falls, or at what speed it reaches the ground, he built a very complicated reasoning to LOGICALLY reconcile this phenomenon with the entire system of his picture about the universe: the famous model "Earth Air Water and Fire". He claimed that heavier bodies fall to the ground sooner than lighter ones because (sigh!) they have a higher percentage of ingredients (Earth and Water) that "are willing" to stay on the ground.
Well, two thousand years later Galileo shows up and performs this simple experiment that you too can repeat at home: take a piece of paper – an envelope is fine - and a book whose dimensions are larger than the piece of paper you have chosen. Kneel on a carpet,

just so as not to make too much noise, hold the book in one hand and the piece of paper in the other, bring them to the same height and let them fall at the same time. It will not surprise you to note that the book hits the ground earlier, just as Aristotle said.

But now pick them up and simply place the piece of paper on top of the book. This piece of paper is NOT glued to the book, it is free to move. Drop the book and oops ... book and piece of paper hit the ground TOGETHER!

Congratulations, you just laid the foundation of classical physics!

After Galileo, Isaac Newton arrives, followed by many others, and in less than two hundred years, classical physics seems to be able to predict practically everything: the motion of rigid objects, - both in space and on Earth- as well as the properties of gases and fluids are now under our full control. Even optical, electrical and magnetic phenomena seem to have no more mysteries. Even chemistry, using Galileo's quantitative approach, manages to discover that we are made of atoms - in 1800, thanks to John Dalton. (This is also a very interesting story to read, maybe in another book).

Only one problem was left without a final word: what is light made of?

Among the many, only two hypotheses had survived: one dating back to Newton himself, who claimed that light was composed by small corpuscles, and another belonging Christian Huygens from the Dutch school, who instead claimed that light was a wave.

Well our journey starts from there.

2.2 - Thomas Young's Experiment - 1800

Newton's theory was the easiest to understand. After all, to our eyes, light seems to travel in a straight line, as in the drawing 2.1 below, where we can see that the size of an object's shadow, projected at various distances, behaves just as if light were propagating in a straight line.

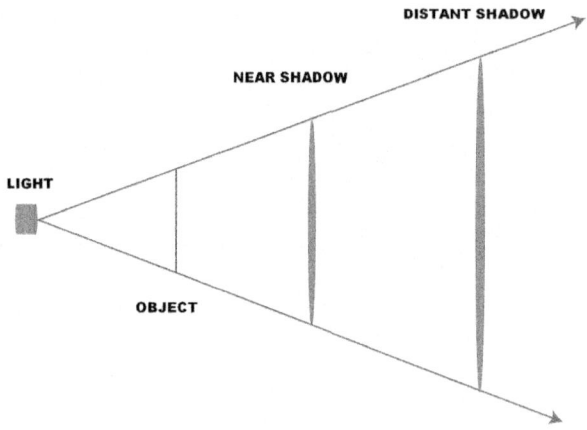

Fig. 2.1

Hence, if we assume that the source fires straight "bullets" of light, it is easy to reconstruct this picture. The same holds true for almost all optical phenomena. In addition, since visible light is made of various colors, we can assume that each color corresponds to a particular kind of bullet and that's it.

Huygens's hypothesis, on the other hand, was a little more complicated to "visualize", and indeed I don't even try. However, with a little mathematics he could manage to obtain the same predictions Newton got with his bullets model.

Like I said, it was more complicated mathematically. Now let's try the following experiment together:

choose two identical or nearly identical objects, for example the index and middle fingers of your hand (or two mobile phones or two DVD cases) and start looking, preferably with just one eye, between them. Make sure to have a bright background, for example in front of a window or perhaps in front of a computer screen lit with a white background in a dark room. Do not stay too far away and slowly bring your fingers together and continue to look in the middle of the slit that is gradually tightening (with DVD cases you can even get them to touch each other, without then pushing them too hard, because some slits still remain unfilled and the whole setup is more stable). Eventually you should start seeing dark strips appearing inside the crack, PARALLEL TO THE SLIT, which alternate with the background image, a bit like in the drawing 2.2 where I overemphasized the effect.

Fig. 2.2

What are these dark streaks? Is my vision defective? No folks, they are real! And I really hope that by now some of you have suspected that they are close relatives to those "entire areas composed of zeros only" that you saw in the image (Fig. 1.9) at the end of chapter 1 relating to waves.

Well, go figure! those dark lines are points in space where light destructively interferes with itself. But how? Where is the second source you told us about?
I said that interference also occurs in other cases, mathematically more difficult to deal with. Well, the case of a small slit (small compared to what?) is conceptually identical, as well as mathematically similar, to the case of the two sources.
This phenomenon is called "wave diffraction". Let's say that a small (again?) slit is like the instance of a reasonably limited but infinite number of sources, one for each point of the slit itself (this is one of the miracles of mathematics, where the concept of infinity can also assume the attribute of large or small).
Even Thomas Young knew the theory of wave diffraction and perhaps he too had the doubt whether those lines were real or an illusion: so he tried to devise a real interference experiment. In doing this he took into account what was suggested by the phenomenon just described, i.e. everything must be very small (here we go again, small compared to what?) otherwise you risk not to see anything, just like with the crack when it widens beyond a certain limit.
Young did not have a laser or other sophisticated tools we have today. Thus, in order to generate the two light sources, in phase with

each other, he took a box and cut two very small and very close slits on it and aimed the box to the sunlight, in order to block all the light, except for the light passing through the two slits. He then made a hole in the box that would allow him to look inside and see what kind of image was created on the side opposite the two slits.
Schematically, as in the drawing 2.3 (not to scale) where we see the box from above, the slits are on the left hand side and the image we must observe is generated on the right hand side.

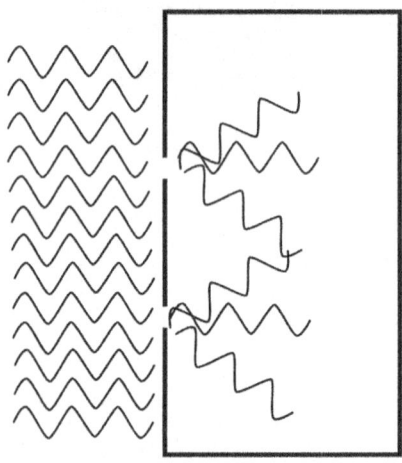

Fig. 2.3

If light were a wave, as I have already drawn, the two slits would behave like two in phase wave sources (see image 2.4 below with real water waves).

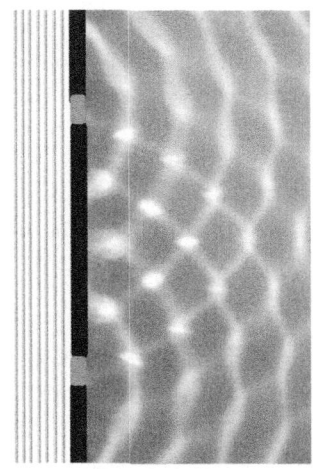

Fig. 2.4

We would see a sequence of shadows and lights. If instead the light were made of bullets, you would see only two bright, maybe a little smeared, spots.

And here's what you see: (*only in some versions of eBook readers).

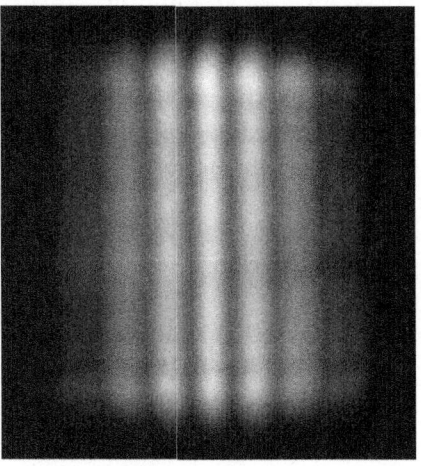

Fig. 2.5

We don't just see alternating light and shadow, as we would expect for waves, but we also see that different colors start spreading out as we move away from the center. What does this mean?

It means that each color corresponds to a different wavelength. In this experiment the central luminous spot is white, because it is in a symmetrical position related to the two slits, so the difference in path between the wave from slit 1 and from slit 2 is always zero, whatever the wavelength is! But as you move away from the center, the path differences are no longer zero. If we express them in terms of respective wavelength, the result is different depending on the color: it will be red if the path difference is an integer multiple of the red wavelength; the resulting light will instead be blue where the path difference is an integer multiple of the blue light wavelength, as so on for all other colors.

The experiment therefore gives us two pieces of information: 1. that visible light (I just introduced "visible" which I had not used until now, soon we will see why) is a wave, and 2. it also allows us to establish the wavelength associated to each color, thanks to the mathematics of interference I mentioned in chapter 1: visible light thus turn out to range between 700 billionths of a meter ("nanometers") for red to 400 billionths of a meter for violet, the shortest one.

Finally, we can see why I insisted on the term "small". Wave mathematics tells us that wave phenomena occur if the size of objects analyzed with the waves are comparable with the wavelength we use (we will see more of this later on). Therefore, we can now understand why the slit has to be pretty tight for us to start seeing the diffraction between the fingers; likewise, Young's two slits have to be narrow and close together to show interference with visible light.

Finally, I suggest you watch the following VIDEO[1], where the experiment is re-proposed with an instrumentation that is as faithful as possible to the original one and where everything we said so far about waves is once again explained (repetita juvant).

2.3 – Electromagnetic Waves

In 1865, the Scottish physicist J.C. Maxwell publishes 4 equations (initially they were 7) which have the power to predict not only all the electrical and magnetic phenomena observed up to then, but also the existence of a "something" which takes the name of "electromagnetic waves". The same equations can also predict the propagation speed of these waves, which appears to coincide with the value that, as early as 1676, had been measured for light. Therefore, when in 1887 physicist Heinrich Hertz conducts an experiment where he proves for the first time the real existence of these electromagnetic waves, the scientific community drops all doubts about the nature of light. Light is a particular form of electromagnetic wave.

End of story. You can go to chapter 3.

For those of you who are curious to understand it more in detail, please keep reading this paragraph and even the next one that focuses on the measurement of the speed of light.

We start from afar, but we'll be fast. The first to hand down descriptions of electrical phenomena are - for a change! -, the ancient Greeks. Indeed, it is from them that the name comes. In fact, they told us that by rubbing objects that were made of amber, which in ancient Greek is called "***Elektron***", those objects became capable of attracting small lightweight pieces of dust or similar. You have seen it too: if you rub a wool sweater, sometimes it attracts your hair, for example.

The same goes for magnetic phenomena: it is said that a Cretan shepherd named "***Magnes***", using his iron-tipped stick, discovered the attraction and repulsion properties of some stones.

Everything remains silent until the eighteenth century, when interest in these phenomena increased and a large number of great minds discovered an infinity of phenomena and wrote many formulas to describe them. Then, precisely in 1865, Maxwell manages to enclose all these formulas into only 4 equations. This is a good thing, because, as I said before, science tries to produce models that describe nature in the most extensive and precise way possible, and also with fewer (and simplest) possible formulas

Maxwell's equations, however, are not so straightforward, thus I don't even write them down. In those equations, Maxwell himself finds out one bizarre (for those times) prediction. These equations predict that if you take an electrically charged body and make it move with non-uniform motion, for example up and down, the electrical and magnetic properties of the surrounding space - properties that from now on we will call ELECTROMAGNETIC FIELD - behave precisely as if they were crossed by a wave. Such wave makes the FIELD change over time, precisely as the example already seen in chapter 1, where the wave traveling along the rope changes the property that we identified with the vertical position of each piece of string. That's not all: according to these equations, these waves can also travel in a vacuum, which was very strange at the time, because we always need a medium to have a wave, like the rope, or the air for sound waves, or water for those of the pond, etc.

Okay, I mentioned this last property just for completeness; let's go back to the wave and the electromagnetic field.

So what is an Electromagnetic Field?

It is the measure of the capacity that each point in space has to generate an electric and / or magnetic force on anything located at that point that happens to be subject to such forces. Just like the gravitational field, which I am sure you have heard of, is the ability of each point in space to generate a gravitational force on a mass that happens to be there. Staying with this analogy, planet Earth generates at any points of space a gravitational field **(G)** whose effect is such that a body of mass **(m)** placed at a certain distance from the Earth (let's say from the center of the Earth) will undergo a force **(F)**, named "gravitational", whose intensity, direction can be predicted from the field's G properties thanks to the simple formula **F=mG**.

The same goes for electrically charged bodies, which generate an electric field, and for those magnetically "active" (don't let me use the term "charged") which generate a magnetic field.

Coming back to Maxwell's equations: they say that an electrically charged object moving up and down will generate a wave, propagating in all directions at the speed of light and altering the electromagnetic field in every point of space it eventually crosses. This is exactly how radio, television, your mobile phones, wifi routers, bluetooth and everything that today is called wireless work. It is necessary to make sure (with various methods) that in an object where electrons (which are charged bodies) can move freely, as in any piece of metal (which we then call antenna), a particular motion is triggered. The type of motion depends on the signal we want to transmit: what originates and travels in every direction is a particular form of electromagnetic wave, which depends on how we make these electrons move in the antenna. If somewhere else we have another piece of metal (we call also this one an antenna, which, as I said before, is a piece of metal with free to move electrons), when the wave arrives, its electrons will be in an electromagnetic field that changes over time. Thus they will begin to feel the new field and move like those we had moved in the "transmitting" antenna: this means that the signal has been picked up.

The first experiment that managed to verify this phenomenon was performed by Heinrich Hertz in 1887, with an apparatus that I depict in a very stylized way and with creative freedom, in the drawing 2.6.

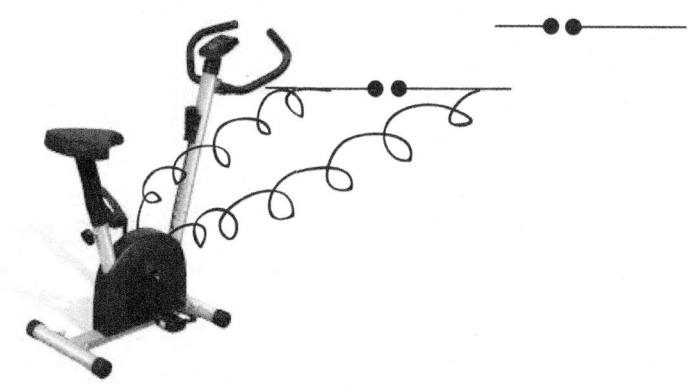

Fig. 2.6

There are two metal rods very close to each other, connected with metal wires to an exercise bike, and ending with metal balls at their ends. Obviously Hertz did not have the exercise bike, but I only need it to show you something with which you can charge the metal rods and therefore also the metal balls.

I don't think I will shock you if I say that by dint of charging, sooner or later a spark strikes between the two balls. This is similar to what we all have experienced in very dry weather, getting out of our car after a long drive and touching the door, only to receive... a shock.

The shock is from the electric charge that suddenly moves, a non-uniform motion. Thus, according to Maxwell's equations, this must generate a wave.

The other pair of metal rods with a sphere, is not connected to anything. If the wave existed, as Maxwell says, something should happen after the spark strikes between the two spheres connected to the bike. And indeed Hertz observes a spark between the other two spheres as well.
The real device used by Hertz can be found here[im1].

Finally, we can conclude that:
• electromagnetic waves propagate in a vacuum, just like light does
• they travel at the same speed of light
• light has also proved to be a wave

So, light is nothing more than a particular electromagnetic wave: case closed!

2.4 – The Speed of Light

As I said earlier, this is an ancillary paragraph, written for those curious to know also how the speed of light was measured. Those who are not interested can easily jump to chapter 3 without the risk of losing the thread.
I'll be brief.
The first one to have the suspicion that light could have a finite speed was Galileo himself, who also devised an attempt to measure it. Two people equipped with a powerful lantern covered by a veil stand facing each other at a certain distance; one of the two removes the veil and records the time; as soon as the other sees the light he too removes the veil, and as soon as the first sees the second flash he stops the time count. The speed will be given by dividing the time measured by distance x 2 (because the light has to go back and forth).
Unfortunately, however, the speed of light is so elevated that the result of this measurement will be affected by the physiological reaction times of the two subjects (and who knows what Galileo had as an available chronometer!!). But it was a beginning.
In 1676 the Danish astronomer Ole Rømer, instead, used the astronomical observations in progress at that time, which were dedicated to the revolution around Jupiter of "Io", one of its moons.

He realized that when the Earth was close to Jupiter, the moon "**Io**" turned out to become visible from the Earth several minutes earlier than the predicted time, which instead was correct when the Earth was far from Jupiter.

Let's look at this very simplified drawing (fig 2.7), not to scale, etc.

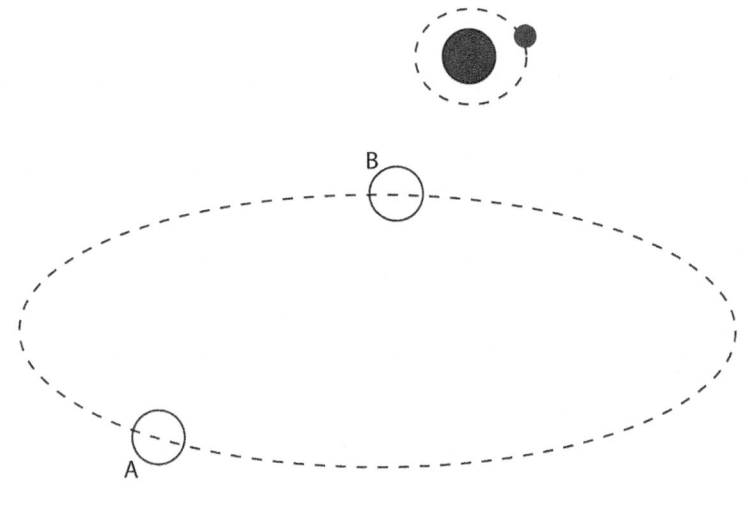

Fig. 2.7

We see the Earth in two different positions (A and B) along its elliptical orbit; we see Jupiter and the moon Io revolving around it. The drawing tries to capture the very instant the moon becomes visible from Earth. If the speed of light is not infinite, then it will take some time BEFORE the image of the moon reaches the Earth, and there will be a difference if I am in position A rather than B. So, if I compare the two measurements of time at A and B and I know the distance between A and B I can calculate the speed of light. This result is very close to what is accepted today, especially after the refinement of astronomical measurements that took place over the years. So already in 1800 we had an almost "correct" value.
Finally, in 1851, the French astronomer Hippolyte Fizeau perfected Galileo's idea and came to a result, which was practically exact. The problem with Galileo's method was in the measurement of travel times that were subject to human reaction times. Fizeau's idea was the following: I take only one lantern and instead of the second, I place a mirror somewhere afar. Then I place a toothed wheel near the lantern, the rotation speed of which I can vary at will. The light

from the lantern begins to travel towards the mirror only when it passes through one of the slots in the wheel, then reaches the mirror and comes back. However, if the speed of rotation of the wheel is such that when light has completed its round trip, the next tooth is in the place of the slot, I do not see the reflected light because it remains behind the tooth of the disc. By knowing the speed of the wheel, however, I know how long it takes for the tooth to completely replace the slot, and that's it.

You will find an interesting rerun of the experiment in this VIDEO[2] and I invite you to reflect that for Fizeau it was much more complicated, because he did not have lasers and not even instruments as precise as today's tools.

Chapter 3. End of the Century – Mother Nature Strikes Back

We have reached the end of the nineteenth century: the most recent successes of classical physics, as seen in the previous chapter, in addition to all those already obtained in the previous years, seem to have lead a physics' giant of that century, Lord Kelvin, to utter such a phrase " *There is nothing new to be discovered in physics now. All that remains is more and more precise measurement*"!
Mother Nature, however, had finally decided to reveal to us two secrets that She jealously had kept hidden, thus forcing physicists to lay down the foundations of what later became Quantum Mechanics. Such secrets were the ultraviolet catastrophe and the photoelectric effect.

3.1 – Photoelectric Effect: the Problem

I decided to start with the photoelectric effect even if the two stories are almost contemporary, because I think it allows us to better see the logical thread that followed.
Let's go back to 1887, in Heinrich Hertz's laboratory. For our purposes I novelize the story and simplify it, but without cheating. Hertz notices that a zinc plate, previously charged in some way, loses its charge when exposed to a strong light source. For those interested in reading how these experimental evidences could be seen at that time, I refer you to paragraph 3.2.
Hertz's result was perfectly compatible with everything we said so far. Just for curiosity, in those days it was thought that the electric charge was a certain "thing" that was loaded on the bodies (hence the name "charge"). The fact that a wave (visible light) swept it away the faster as more intense was the wave, did nothing but confirming this view. This vision remains valid even when, a few years later, with the discovery of the electron, the model is changed into

assuming that the electric charge is due to an excess or deficiency of electrons.

Let's go back to the experiment. If a transparent glass plate was placed between the zinc and the source, the process did not take place, even if they doubled or increased tenfold or more the light source intensity.

This is a surprise! However, we should remember that glass is transparent only to visible light, while it blocks the ultraviolet "colors".

So maybe ultraviolet frequencies are the sole responsible for the zinc plate discharge!

The phenomenon is then investigated by many others and different metals are tried over the years.

To test the effect of the various colors, colored glass is used and the result is that, for some metals, the discharge occurs also with normal glass or with blue glass, but not with glasses of other colors whose frequency is lower than blue.

By using other metals, it turns out that one can go down to the green color. I wrote "go down" because green has a lower frequency than blue. With some other metals it is even possible to go down to yellow. For every metal, however, there is this anomaly: if one lightens the metal with a "color of a frequency which is too low" (we have seen that for zinc this threshold is in the ultraviolet region) the metal does not discharge, no matter how strong the light source is.

However, the mathematics of waves, any wave, and therefore also of light (if light is a wave) establishes that energy is linked to its amplitude or intensity, certainly not to its frequency.

I think it is evident even to us. Let's think of a bucket filled with stones placed on the beach. If many small waves arrive in a minute, the bucket doesn't move. If in the same minute I get even a single wave two meters high, the bucket certainly moves!

The solution (Par 3.4) will be discovered in 1905 by a "certain" Albert Einstein, who drew his inspiration from what I am about to tell you in paragraph 3.3.

The next paragraph focuses instead on showing you how they could observe these things back then, therefore, it does not serve the thread of the story, but since the tools that were used can be found today in any home, I think it may be interesting for some of you; moreover, you will also find other curiosities about physics that are not directly connected to this story but that could be worth reading. E.g., answering the question: "what is the electric charge"?

I repeat, those who are not interested in "getting their hands a little dirty" can jump directly to paragraph 3.3.

3.2 – Hertz Experiment in your Kitchen

Let us first try to understand what the electric charge is from a down to earth viewpoint. Have you ever seen one? And how can we say that there are both positive and negative charges?
Let's proceed in order.
We have seen that if we rub a wool sweater, or plastic, or polystyrene, these take on the ability to attract light objects (hair, cotton balls, pieces of paper). Now you must know that when physicists resumed studying these phenomena in the 1700s, Newton's law of Universal Gravitation was already in place.
But what does Gravitation have to do with it?
Well, Newton's law was the first formula that describes the behavior of "force at a distance". I say "force at a distance" precisely because gravity acts without touching things. The "electric" force does that too. Let's go back to the law of gravitation: this establishes that the intensity of the attraction force between two bodies (for example the Earth and us, but also between us and the book you are reading) becomes bigger and bigger as a property of each of them increases - precisely their mass -, and decreases as their distance increases. The "distance" effect also seems to apply to the electric force; it was therefore assumed by similarity that there is a new property of bodies, which we call charge, that generates an electric force just like mass generates gravity. The big difference is that while mass is a clear and visible property of bodies, you don't see this new property we call electric charge.
And again, how do we say that there is a positive and a negative charge?
Well, another peculiarity of the electric force is that it is attractive but also repulsive.
If you want to see the repulsion, try the following experiment: put on a wool sweater and take two fruit department gloves from the supermarket. Stand on a carpet or wear rubber shoes. Rub one of the two gloves on the sweater you are wearing. The glove will stick to the sweater. Do the same with the other glove. You now have two gloves attached to the sweater. We are now at the most delicate step,

which is not always successful because when we are rubbing objects, there is a big risk that even your arms and hands are now no longer electrically neutral.

I therefore suggest you move slowly and touch a large metal object, such as a floor lamp, or the fridge, or a heater, before proceeding any further. Now, slowly, take one glove with one hand, preferably from the glove's middle finger, do the same with the other and keep your arms outstretched (everything should stay as far as possible from your body, which as I said may not be neutral) and then bring your two hands close to each other. You will notice that the gloves do not attract each other at all; on the contrary they tend to repel each other forming - if they hang from their middle finger-, a sort of inverted V. Don't worry if it does not work out right at the first attempt; try again and you will see that sooner or later you succeed (see photo 3.1).

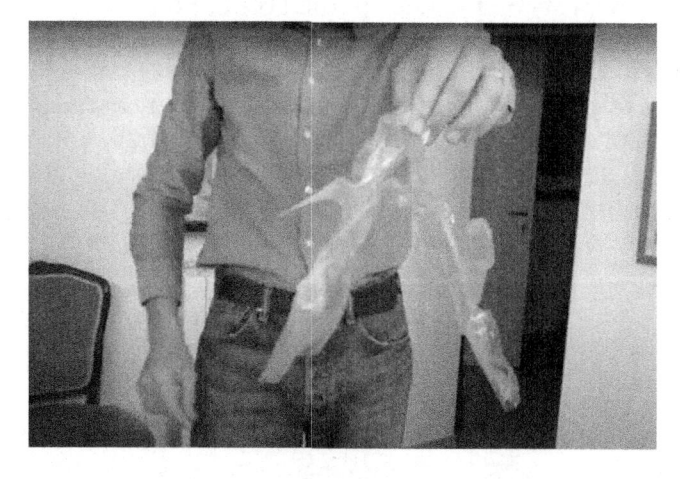

Fig. 3.1

This is how the concept of positive and negative charge was born. We are forced to assume that there are two types of charge: when they are of the same type, as in the case of the two gloves, the force is repulsive, and when they are opposite, the force is attractive. Later on, Benjamin Franklin decided a priori which one was to be considered positive and which negative and from there all the other charges are determined on the basis of their behavior with respect to the "sample" charge.

Another thing that was noticed was that one cannot charge metals just by rubbing them (try and see), although metals seem to have the property of "transmitting" electric charge, just like they do with heat. Indeed, if you approach one end of a long metal body with an object that you have previously charged, you will notice that electrical forces occur at it's far end, but they disappear as soon as you move the charged object away.

To better understand what I just said, try this test, which actually allows you to build an instrument that we will use shortly, and which is called an "electroscope". Take a big plastic bottle and a piece of cardboard that you can place on top of the bottle. Then take some metal (iron or copper) wire, a bit stiff for convenience (a television antenna cable is fine), fold it into a hook on one end and slip the other end through the cardboard, making sure the cardboard will hold it. Then take some tinfoil: wrap a good piece of it around the tip of the wire that you have not folded, while on the hook you hang two small thin sheets, also made of tinfoil. Now carefully thread everything into the bottle from the hook side until the cardboard rests on the neck of the bottle. You will therefore have a good portion of the wire inside the bottle, not touching the bottom, while the tip of the wire and the tinfoil ball are outside. Now take a good piece of polystyrene, rub it well (so that you charge it) with the usual piece of wool and then bring it close to the ball. You will notice that the two small foil sheets inside the bottle move away from each other, forming the inverted V we've seen with the supermarket gloves, as shown in the photo 3.2.

Fig. 3.2

What is going on here?
The charged body approaches the metallic one. But in metal objects charges are free to move, so the charges opposite to the charged body will be attracted to it and will concentrate towards the ball, while the other ones will be rejected and pushed towards the leaflets, which are free to swing and, having identical charge, will behave exactly like the gloves. You have created a rudimentary electroscope, which allows you to understand if an object is charged, although it does not allow you to understand its charge sign. Okay, you can't have everything in life ...

Now all that's left to do is charging a metal object and bringing it closer to the electroscope to see what happens if we shine the metal object with a light. However, charging a metal object isn't that simple. We have seen that nothing happens if we rub it.

Here we must resort to Alessandro Volta's genius. Let's see what he invented:

We have just seen that if I approach a long piece of metal, or more generally a conductor, with a charged object, the charges inside the conductor will drift apart: the charges of a specific sign move to one tip and the charges of opposite sign to the other. In Volta's days, it was already clear that even the human body appears to behave like a conductor: they used to play lots of "pranks" where an accomplice was hung horizontally, someone or something vigorously rubbed the woolen socks worn by the accomplice and then the tricked victim was asked to give a kiss to the accomplice and boom ... a tremendous shock. So let's take a large object that we charged by rubbing it, let's place it on a wooden table, perhaps even on top of some piece of wool, let's take a metal object and place it on the charged object and let's keep touching it. Well, metal is conductive and so am I, so the charges opposite to those of the charged object are all in the metal piece that we are holding, while the opposite ones are at our feet (it is even better if we are barefoot on the ground). Now we detach our hand and the metal remains with only one kind of charge.

From this moment on we have to make sure we handle the metal using insulators to prevent it from discharging by the new contact with us.

Ok now you have everything: you have the charged metal, you have the electroscope to measure if it is charged, and you can have fun experiencing Hertz's experience of the photoelectric effect.

3.3 – Planck and the Ultraviolet Catastrophe.

Today, when you buy a light bulb, the associated "color temperature" is shown, amongst other features. What is this color temperature?
It stems from the evidence that every object emits light (i.e. electromagnetic waves) and the type of light depends on the temperature of the object. We emit light too, but our temperature of 36 °C (96,8°F) makes it invisible to our eyes, so if we are in a room with no lights, no human being can see us.
However, anyone who has followed any TV reality show in recent years has surely had the opportunity to notice that an infrared light sensitive camera can see us, even in total darkness.
We don't need the infrared camera to experience the temperature-color connection; for example, when you heat a piece of metal you notice that as the temperature rises its color changes. The old filament bulbs worked just like that. Moreover, if you look at the metal that makes up a top grill of any oven you will see how the color changes as the temperature rises.
Back then, physicists already noticed this dependence of color on temperature and devised a way to quantitatively measure color at different temperatures.
How do you measure color quantitatively?
It was already known that the perceived color is the result of the mix of visible colors, and depending on whether there is more blue or more red, white light looks different (we use the term "cold light" or "warm light").
Why bother knowing the ratio among blue and the other colors in the light emitted at various temperatures?
If we want to try to infer the temperatures of various stars, for example, it will be of great help to start from their colors "signature".
I take a telescope, aim it exclusively at the star I want to analyze and direct the light that comes out of that telescope to a grid (called a diffraction grating) similar to that used by Young (Par.2.2), but with many more slits, in order to have a better spatial separation of the various colors. Previously, I placed thermometers where I expect to find each color and by comparing the increase in temperature of each thermometer during a given period of time (for example an

hour) I can tell how much red there is compared to how much yellow, how much green, how much blue etc. (see drawing 3.3 below).

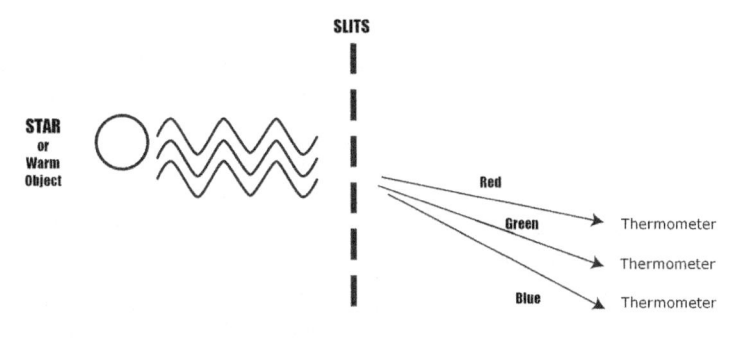

Fig. 3.3

If I do this very same exercise using light emitted at various known temperatures by ordinary objects, preferably a black object because it doesn't reflect any light (and in fact this is called "blackbody radiation", but this is not so important), I obtain a nice series of color signatures of all temperatures and I can make a comparison with the signature of the star.
Everything works perfectly well, except for one small detail. Physicists do not like having to deal with empirical tables: they want to have a theory that predicts the values of the tables; thus they try to use existing formulas of the theories at their disposal, i.e. those of classical physics.

But unfortunately nobody seems to succeed. Whatever the procedure used, in the end some formulas are obtained which, given an object's temperature, manage to predict with good accuracy only the portion of colors of low frequencies, let's say up to yellow (so to speak). On the other hand, for colors beyond yellow, the predictions of the formulas differ considerably from the experimental data. More specifically, the formulas that are produced by the various attempts have a big problem in common: they all foresee that the "quantity" of a certain color grows with its frequency, regardless of the temperature. This does not match the experimental data where in reality starting from a certain frequency onwards the contribution decreases. As you can guess, the theoretical result is also absurd, because it would mean that I will always have an infinite amount of light emitted, given that the possible frequencies are infinite.

Infinite amount of light means having infinite energy and this is not possible. This phenomenon takes the name of "ultraviolet catastrophe" precisely because the error of the theory occurs for high frequencies, while for low ones it all adds up.

The solution is found by physicist Max Planck and it is not easily understandable. It can only be reached through complex mathematical calculations and so this time you will have to trust what I say.

Planck notices that the formulas derived from the theory come perfectly in line with the experimental data, for any frequency and at any temperature, as long as the following idea is introduced: the energy associated with any electromagnetic wave (light) of a given frequency **"f"** cannot assume just any value, but only integer multiples of a quantity hf where "f" is the frequency and **"h"** a constant, identical for all, which today is called Planck's constant.

In other words, a wave of frequency f can have energy equal to **0, hf, 2hf, 3hf... 1000hf but not (for example) 3.4hf** or so.

This was a big departure from Maxwell's theory, which did not offer such discretization, or I should say quantization, of electromagnetic waves energy.

And further: this result emerges only from mathematical considerations, but it was not clear what physical phenomenon could be behind it. Planck himself was skeptical about the validity of this invention, but today the formula **E = hf** (or better **E = hν**, where **"ν"** is the Greek letter "nu" or "ni") is the basis of quantum mechanics and of all the theories deriving from it.

The first guy who took Planck seriously was Albert Einstein in 1905, who used this hypothesis to brilliantly solve the problem of the photoelectric effect.

3.4 – Photoelectric Effect: the Solution

We have said that only a certain fellow named Einstein took Planck's conclusions seriously.

Let's go back to the photoelectric effect. We now know that we cannot succeed in discharging a piece of metal, especially if negatively charged, if we use a light whose frequency is below a certain threshold, even with very high intensities. On the other hand, if we use a color light that exceeds this threshold (whose value depends on the type of metal we are using) everything proceeds as expected: we discharge the piece of metal, and the more intense the light we use, the quicker the process is.

Einstein claims the following idea: we know that negatively charged metals have an excess of electrons on them (the electron was discovered less than ten years earlier); let's assume that each metal exerts its own small bonding force on these excess electrons.

Now, ATTENTION TO WHAT YOU ARE ABOUT TO READ, let's assume that light (or rather any electromagnetic wave) is not a wave in the common sense of the term, but is composed of many small bullets (which today we call "photons") that always travel in a straight line at the speed of light (no wonder!!). Let's assume photons have zero mass, are point-like but nevertheless each of them carries an energy equal to $E = hf$, exactly as Plank says, where f is the frequency of the wave we are dealing with. I repeat, the photons travel in a straight line, just as we see in a ray of light, they are not bumping up and down. With these assumptions we can say that the photons hit the metal, one at a time. If the wave is very intense, there will be many photons one after the other, while, if the wave is not very intense, there will be few, but always one after the other.

Each photon therefore has the possibility to hit only one of the excess electrons and to supply it with all its energy hf. But if this energy, which I repeat amounts to hf, is not strong enough to break that slight bond the metal exerts on every single electron, the struck electron remains bound to the metal, and everything happens so quickly that even if the same electron is hit by a second photon nothing happens. Therefore, no matter how intense our light is, or rather how many photons arrive in a certain time, none of them can undermine any electrons and the metal does not discharge. Conversely, if the frequency of the incident light, and therefore of each photon, is such that hf exceeds that bond between electron and metal, then each photon is able to undermine the electron they hit and the metal is discharged, all the more rapidly the more intense the wave is, that is, the more photons arrive in a certain period of time.

Needless to say, Einstein's arguments are not just qualitative, but they turn out to match experimental data, thus this paper is a big success.

Hence, the wave-particle duality is born, and physicists must force themselves to consider light (just light for now) as something that behaves both as a wave and as an array of bullets (photons).

Please be advised that this new vision turns out to replicate EVEN the experimental data that were predicted by Maxwell's theory. This is why it is " a better theory".

But the story doesn't end there, as we'll see in the next chapter.

Chapter 4. De Broglie (1924) – The Wave in You

4.1 - De Broglie's Ph.D Thesis

After Einstein's launching of this wave-particle duality of light in 1905, nobody (Einstein included) has a clear picture of what that actually means, but it works very well. Other experiments, like Compton's effect (I will not talk about it), confirm the validity of this novelty.
Let's jump forward to 1924; later on, in the next chapter, we will come backwards to the terrible 1887.
In 1924, in Paris, a student named Louis De Broglie is taking his final doctoral exam. In his thesis, he offers a bizarre vision, which is more or less the following: Mother Nature does not make preferences. If She needs photons to create what we call light, or more generally electromagnetic waves, - which in fact is both wave and matter-, then by symmetry, She must have given a wave nature even for the ingredients She uses to create what we call "matter". In other words, everything (you, who are reading this book, as well as the book, the table, the chair, etc.) must also have a wave nature.
Be careful, he does not mean that you are bouncing up and down. Actually, according to De Broglie, your current state can and must be described not only as that of a material point (just to simplify), as we do in classical physics, with a certain mass, in a certain position and with a certain speed, but also as a wave, with certain properties.
What wave properties are we talking about?
Well, he still cannot envisage all of them, but he has no doubts about a couple: the wave always travels at the speed of light **"c"** regardless of your actual "classical physics" speed **"v"** but in the same direction of your speed, and - again by symmetry with the case of light - has a frequency **"f"** which comes from the well-known Planck formula $E = hf$, i.e. it depends on your "classical" kinetic energy E which is a quantity that depends on your "classical" speed "v" and on your mass **"m"**. In reality, in his formulas, De Broglie does not use kinetic energy "E", but another property which is typical of classical

physics, and which is a function of the classical speed "v" and mass "m" of an object: the "momentum", and he links this momentum to the wavelength, not the frequency. I didn't use the real formula just because I want to stress the similarity with the electromagnetic (the light) case, trust me when I say I'm not cheating.

THE MAIN CONCEPT to grasp is that he argues that even a material object of mass **"m"** and velocity **"v"** behaves like a wave, whose wavelength **"λ"** (I told you in chapter 1 we will use wavelength or frequency at our convenience) is an expression of h, m and v: λ= **h/mv**

Folklore note: the examining commission was about to reject the thesis, were it not for the fact that Einstein, by then a famous scientist, happened to be in the premises and said something like: *"this in not such a bad idea, maybe it's even true"* and so the student got his doctorate and the thesis didn't end up in the shredder.

4.2 - 1927 Davisson & Germer's Experiment

As I have tried to stress since the preface of this book, a scientific theory cannot be based only on the credibility of the person proposing it, or on its mathematical rigor, or on the logic that is behind it.

A new theory, in order to supplant the previous one, must show that it has a greater predictive capacity (more phenomena explained) and / or a better one (more precise). It won't hurt if it turns out to be even simpler, but unfortunately this third feature hardly happens.

Hence, we need experimental verification of De Broglie's idea. Back then, there was no internet, the news did not travel fast and, given the attitude of the doctoral commission, we cannot envisage that De Broglie's thesis ended up on the front pages of the newspapers. We can suppose rather that it remained a kind of gossip wandering between scientific circles (maybe I am exaggerating, just to dramatize).

Next, I am about to describe one of the most striking cases of "Serendipity" of which physics' history is full.

Two American physicists, Davisson and Germer, are doing studies to better understand the geometry of the surface of some metallic "crystals", or rather "crystalline solids". The term "crystal" does not mean the material that composes precious glasses, but those particular pieces of rock that are composed almost entirely of a single element: we could call them "nuggets" if they were metals, as was the case in question, where the object of the analysis were nickel "nuggets" (figure 4.1).

Fig. 4.1

You should know that for quite some time by then, physicists had learned a trick to help them gain additional information on the "shape of objects", compared to what could be obtained by just observing them, even with the most powerful microscopes. This method consisted in firing bullets, the smallest ones possible, onto the sample to be analyzed, and measuring how they scattered off the target.
Let me give you a rough example using the drawing 4.2 below:

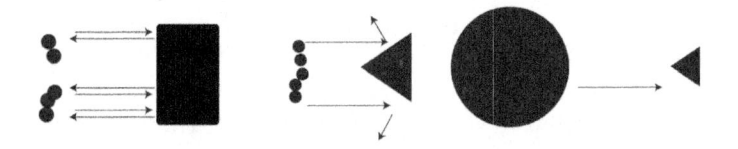

Fig. 4.2

It should be intuitive, and even more evident from the drawing, that the scattering result of the individual bullets off the target allows us to guess the target shape. This, provided that the dimensions of the bullets are not equal to those of the target, in which case (last drawing on the right) I think it is obvious that the outcome of the collision is totally independent of the shape of the target.

Even more so, if I want to analyze the details of each part of the target, I would have to use the smallest bullets possible. Furthermore, classical physics allows us to deduce the aforementioned details in a quantitative and precise way, as long as we both know the initial conditions of the bullets and we can measure their final conditions.

Davisson & Germer, for better understanding the details of the surfaces of these nuggets, decide to use the smallest bullets available at the time, namely the electrons (which, I recall, had been discovered towards the end of the previous century).

Basically, they create the thinnest possible beam of electrons, point it towards different areas of the nugget's surface, and observe where they scatter, as you can see (very schematically) in image 4.3 .

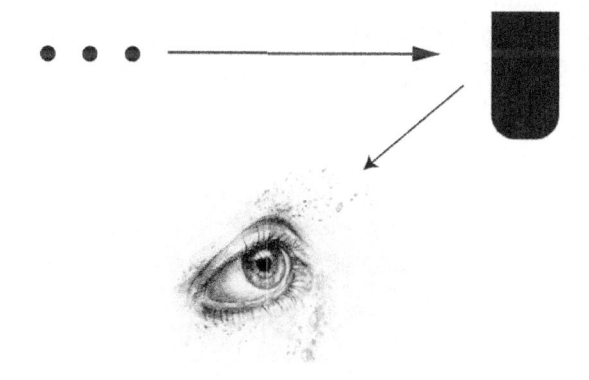

Fig. 4.3

Suddenly, the news of the bizarre De Broglie's wave theory reaches them. So what?

Well, take a look at the picture 4.4 below, which you can also produce at home. Some light is reflected off the surface of a very common television set...

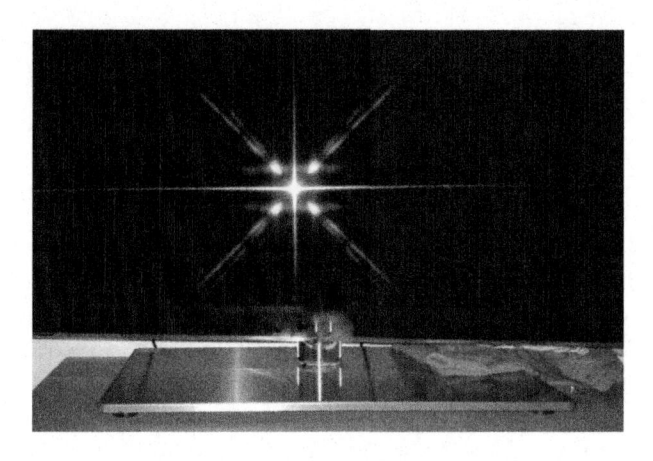

Fig. 4.4

... and on the diagonals you will see that a series of rainbows* **(*colors are visible only in some versions of eBook readers)** are formed, in short, a figure similar to that of Young's experiment. You'll get the same result if you look at the light reflecting off a CD. If you move the CD in various ways, sooner or later, you will see the different colors of light appear on its surface.

What is happening to the light, both in the case of the TV and in the case of the CD? Well, the surfaces of the TV and the CD are not completely smooth, but they also have grooves - or rather slits - and we can guess that the distance between these slits in certain directions is such as to cause a diffraction (or interference if you please) phenomenon to the reflected light.

Once again: So What?

Well then, someone reminds Davisson and Germer that even the smoothest surface is not exactly smooth, because it can (indeed must) be depicted like a thin layer of atoms, as in the drawing 4.5 below (just the left surface for simplicity).

Fig. 4.5

We therefore once again have an array of slits that in the jargon is called a "diffraction grating".

And further: in those days, there existed already an estimate of the size of atoms and, as a consequence, of the spacing between slits created in a layer of atoms. Needless to say, it turns out that the size of this spacing is theoretically compatible with the size that the wavelength of the electrons would have, if De Broglie's theory were true. For this reason, Davisson and Germer understand that they should be able to observe an interference phenomenon (or diffraction, at your convenience).

So what do Davisson and Germer do? They place themselves at a certain angle to their sample and start to measure how many electrons arrive at that angle by varying the energy of the incident electrons. I remind you that, by varying the bullet's energy, the frequency, therefore the wavelength, of this "phantom" wave varies, according to De Broglie's formula; if I place myself at a certain angle, I will notice a sequence of high and low rate of electrons (constructive/destructive wave interference) corresponding to the change of the incident energy.

In the figure 4.6 is the result of their measurements, where along the horizontal axis we have the energy and along the vertical axis the electron count over a certain period of time.

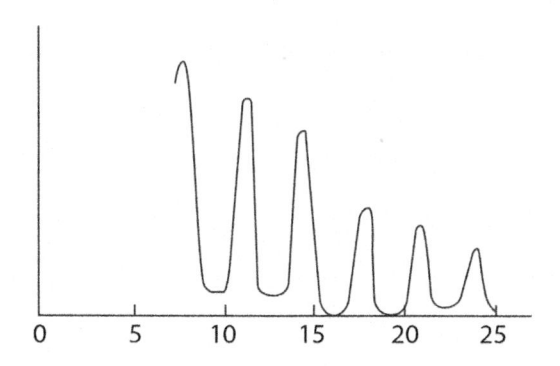

Fig. 4.6

The measurements confirm both the wave nature of the electrons, and quantitatively the correctness of the De Broglie formula, which links the electron's wavelength to its mass and speed.

Conclusion: There is no clear-cut distinction between wave and bullet. All the basic ingredients of our universe, therefore also their products, including you and me, have this wave particle duality. In my case and yours (and the book, etc.), we cannot see this wave nature because, just as the formula says, the wavelength associated with us or with the table, etc. is so short that the objects we normally interact with cannot reveal the wave nature. In this regard, I remind you that in order to see the diffraction between the fingers, we have to bring them very close to each other, otherwise if we separate them too much, light will behave like rays going straight as if they were bullets.

Now the question is, what the heck is this wave?

Chapter 5. Born, Heisenberg, Schrödinger & His Cat

The chapter you are about to tackle will be the most difficult to read. The problem lies in the fact that absurd situations will arise, even as the outcome of a logical reasoning. Precisely, the interpretation of what the wave is (incidentally we will refer to what is known as the "Copenhagen Interpretation") forces physicists to abandon the link we have with the perception of the world in our daily life and to accept situations that for our brain appear impossible.

5.1 - Max Born

Although, in the previous chapter, I wrote that De Broglie's hypothesis had not appeared on the front pages of newspapers, actually, in those days, physicists were sort of rock stars and the newspapers were usually very concerned with their discoveries. So some physicists did become passionate about this subject and, among these, we mention Max Born.
In 1926, that is even before the experimental confirmation by Davisson and Germer, he tried to reconcile the wave aspect of the electrons (on which he therefore placed a "faithful" trust) with their "corpuscular obstinacy". What do I mean with this expression? I mean that any experiment that involved electrons and recorded their position showed that every single electron left a point-like trace on the detector (regardless of the kind of detector used). There

was never a case of an electron leaving a random smeared trace, as it would be expected for a wave.

Only Born manages to express a hypothesis that potentially provides a conciliation between these two aspects. He proposes that the wave does not represent the arrangement of each fraction of an electron in any point of space; this was the main interpretation running, that implied that, for some mysterious reason, each electron would have transformed from a point-like object into something scattered everywhere and crossed by a wave like a liquid or sand. Born proposes that this wave represents the probability that an electron (therefore also a photon) can be found, AS A WHOLE, in a certain position.

In other words, Born suggests that the De Broglie's wave associated with a moving electron (but, I repeat, it also applies to the photon), does not represent, in a certain instant, the distribution in space of each portion of that electron, but the probability to find it, at that very instant, AS A WHOLE and in a POINTLIKE SHAPE in that position. You should already see how this concept clashes with our natural concept of the trajectory of an object. For us, at any moment in time an object has a well-defined speed and position. Here, on the other hand, it seems that if the object's velocity is known, then its position is not defined, but there is a probability other than zero (I repeat, in the same instant) that the object is in one of the infinite possible positions where the value of the wave is not zero (actually it's even worse than that).

I hope the next paragraph will help you digest this nonsense better, because I am about to give you the definitive experimental proof (we already have Davisson and Germer).

5.2 – The Double Slit Once Again (1961)

Let's move forward in history for a second.

Following Davisson and Germer's result, the scientific community was working to replicate the experiment, but this time with a "cleaner" double slit experiment we have already seen with Thomas Young.

However, only in 1961, physicists managed to carry out such an experiment for the first time. You can find extensive documentation

of this online: just type "double electron slit" and you will see many of the pictures I also show here, as well as interesting videos.
This is the reason why I only quickly describe the experiment, while I focus mostly on its interpretation.
Schematically -see drawing 5.1 below-, we have the usual electron gun, an obstacle with only two slits, and a photographic plate which the electrons impress with a small point-like spot after passing through one (are we sure?) of the two slits.

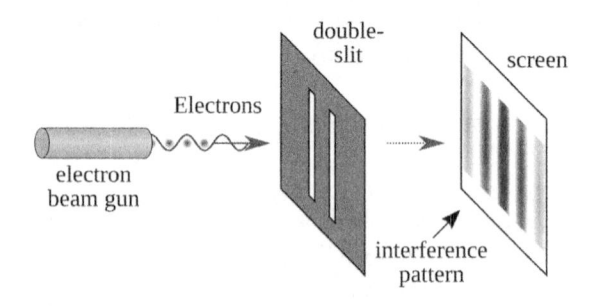

Fig. 5.1

As I said before, each electron leaves a single spot on the photographic plate; but if we wait for a sufficient amount of time, we will start seeing a pattern as in the drawing 5.1 above, also reported in the image 5.2 below (photo "a" after 10 electrons and then gradually increasing).

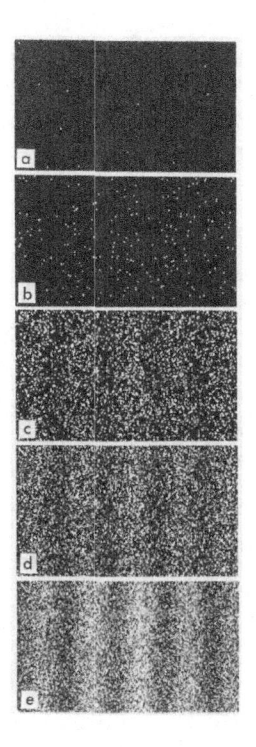

Fig. 5.2

There are areas with greater intensity of hits (electrons) and areas with less intensity, just as what we would expect if a wave went through (just between us, using the formulas and experimental data, the wave properties turn out to match De Broglie's wave). But now, let's look at figure 5.3 where we compare the results we would expect with waves, the result we would expect with bullets and the results observed with electrons.

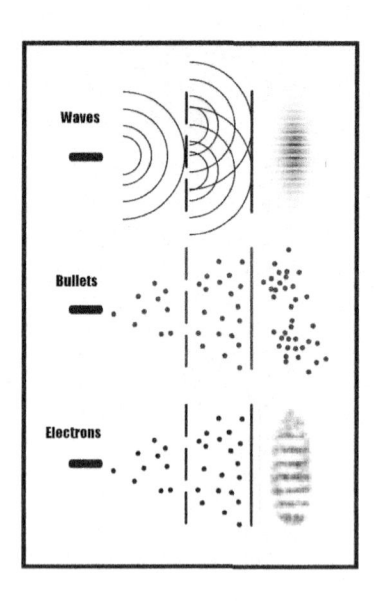

Fig. 5.3

The bullets would form only two distinct large areas of hits, although perhaps partially overlapping, but certainly not with a maximum right in the center. Both the wave and the electrons scenarios show a pattern of maximum / minimum and one of the maximums is right in the center. In other words, the wave intensity on each point of the plate represents the probability of finding an electron in any specific point of the plate. Thus, it should be clear that, this same figure would be recreated by the electrons, once a high number of them has hit the plate, because they are distributed exactly like predicted by the wave impressed on the plate. We can experience this matching of a multitude of identical events to their probability distribution by counting the occurrences of numbers 2,3,..,7,..,12, when throwing two dice many-many times.

In the dice case, the probability of obtaining a 2 (worth 1/36) or obtaining a 7 (worth 6/36), or any other number, is easily calculated, we do not need De Broglie's wave. When I roll the dice, one and only one number comes up, from 2 to 12. But if I roll 1,000 or more times, the frequency of 7s will be much higher than that of 2s, and such frequency will follow that very probability distribution (1/36 for number 2, 6/36 for number 7 and so forth).

In the case of the electron (or photon), the wave travels smoothly, describes the probability of finding the particle (I stop writing "and the photon" every time) in each point of space (attention, both along the vertical and the horizontal axis, but we now focus on the vertical). Then the two slits break the original wave into two waves, each of which keeps describing a probability of finding the particle. However, the two waves, like all waves, interfere with each other, therefore the total probability will be given by the interference pattern of the probabilities represented by the two waves. As a result, the alternation of maxima and minima starting from the central maximum is generated.

As proof of this, if I block one of the two slits, I will have the contribution of only one wave, and the figure that I expect to obtain (and that I obtain) on the plate will be one large area of hits, which perfectly coincides with the figure (fig 5.4) generated by only one of the two waves.

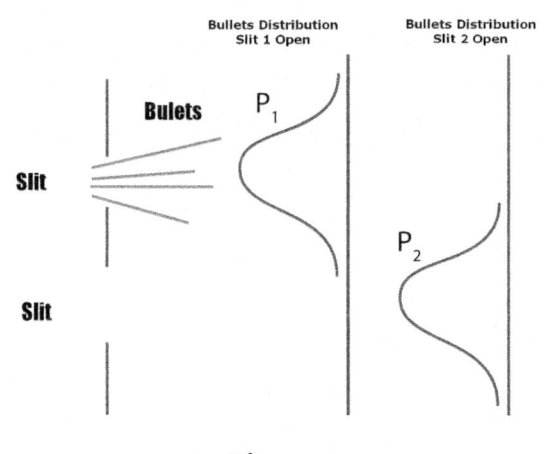

Fig. 5.4

However, there is a much more disturbing aspect that you need to know. Physicists try to modify the experiment in such a way as to leave both slits open, and at the same time they find a way to KNOW which slit each electron has passed through, AFTER it passes the slits but BEFORE it hits the plate.

What did they get in this case?

Well, the figure obtained is no longer a pattern of maximums and minimums, but is a figure similar to the two large areas of spots that would be obtained with "normal" bullets. The meaning of this last result is subtle but it is the HEART of Quantum Mechanics: so make yourself comfortable and read the next paragraph.

5.3 - Copenhagen Interpretation (1927)

And now let's go back to 1927!!!
Yes, because the result that I described in the previous paragraph is indeed the confirmation of an interpretation, which had already been formulated in 1927. For the sake of clarity (?!) let's first look at the reasoning based upon the evidence of the experiment. The conclusion matches perfectly the assumptions formulated many years earlier, thanks to the genius of the many physicists who developed this interpretation, but nobody knows how.
What can we say about the destiny of individual particles in the experiment? Let's go back to figure 5.5 (same as 5-3) below and focus on the wave scenario:

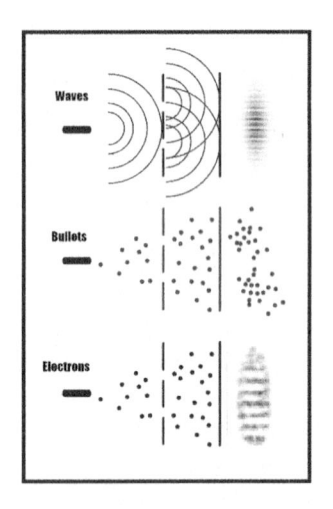

Fig. 5.5

Before passing the slits, we have a single wave, which behaves like a normal wave in a pond. This wave (actually its square value, but let's forget about this), according to the hypothesis we are trying to prove, is describing the probability that the particle is in a certain point of space. Then, the wave reaches the two slits, PASSES THROUGH BOTH OF THEM (and in fact in our experiment we do not know which of the two slits the particle passes through), and two waves are generated, once again exactly as it would happen in a pond. Each wave keeps telling the probability of finding, in a certain position, a particle that passed through the slit that generated that wave. These two waves COEXIST and indeed the waves continue and interfere with each other to give the pattern we see in the "electrons" case. However, if, INSTEAD, for each particle at a certain point in time I HAVE THE CERTAINTY about which slit it passed through, I will also BE SURE it didn't pass through the other one. Thus, if the wave is a probability, from that moment on, one of the two waves (obviously the one originating from the no-pass slit) becomes ZERO; therefore, once again, I have just a single wave that travels up to the plate and produces the bullet scenario pattern. I no longer have the interference because any time I get to know which slit it passed through I destroy one of the two waves. In the real experiment some particles appear to have passed through one slit and some through the other one, so the final image is that of two large blots, like in the case of bullets, where there is no interference but an overlap of the two distinct blots.

In other words, if I don't measure which slit the particle has passed through, the particle behaves as if it had passed through both slits; but if, instead, I manage to measure it, immediately the particle behaves as if it had passed only through that slit.

Let me repeat it again, but from a different "point of view". The behavior of the particle is that of an object that has passed through both slits, unless I have measured which slit it has passed through, in which case the behavior is that of having passed exclusively through that one.

Forgive me if I'm dwelling on this, but it is really the "core" of quantum mechanics. We have come to the conclusion that the particle passes through both slits. However, if I instead get to measure which slit it has passed through (note that I do it after it passes the slits), then it appears that it has only passed through that one.

What you are saying is absurd! The object ALWAYS necessarily passes through one or the other. And, on top of that, assuming you are right, you also say that, if you get to measure which slit, that traces back as if it DID actually pass through one of them only!

I know, it's absurd, but it is as I say. The object does not pass through one or the other at all. Only IF and WHEN I measure which slit it has passed through, the properties of the object (in quantum mechanics we say "the state") change, taking on a precise value regarding which slit it went through. If I don't measure it, this "property" remains indefinite. When I say "indefinite", I do not mean that it is definite and I just don't know what it is. No, it's just indefinite! In fact, the interference pattern proves that there is no exclusive passage from one of the two slits.

What about when you do measure it?

Well, the conclusion reached was that the act of measurement forces the object to choose one of the possible (in this case just two) scenarios, and from that moment on it behaves as if that "choice" had ALWAYS been defined.

This conclusion is called Copenhagen interpretation, because it was the result of meetings among physicists organized in the Danish capital by the physicist Niels Bohr (and which encountered bitter skepticism, for example from Einstein). This Copenhagen interpretation argues that there is no meaning in asking what the value of any quantity of a physical system is before this is measured (any quantity, not only the position and /or the speed). It is the measurement process itself that "randomly" assigns one of the probability values predicted by the wave (in jargon "wave function") describing the state of the system at that precise moment in time.

In the next paragraph, we further explore this latter aspect.

5.4 - Heisenberg & Schrödinger's Cat

Let's try to face two of the most famous "statements" of quantum mechanics, namely Heisenberg's Uncertainty Principle and Schrödinger's Cat Paradox. While reading this book you have already experienced both of them but you probably didn't notice it (because I didn't point them out!).

Both Heisenberg and Schrödinger are active central characters of these Copenhagen meetings and are among the fathers of quantum mechanics. Schrödinger in 1926 had published his famous (?) Equation, which replaces Newton's equation (F = ma). Newton's equation is the cornerstone of classical physics, while Schrödinger's is the cornerstone of quantum mechanics, precisely because it allows us to predict the effects on the properties of this "damned" wave function, due to any forces acting on a physical system, whose "state" is described by the wave function.

Heisenberg was enthusiastic about the results, but Schrödinger was rather skeptical. Heisenberg's enthusiasm manifests itself precisely in his uncertainty principle. In fact, the Copenhagen interpretation, as I have just outlined, does not seem to exclude the possibility that we can still measure ALL the variables that characterize a physical system at a certain moment. However, Heisenberg says: NO. There are pairs of variables that can never be measured simultaneously with absolute precision (in jargon they are called "conjugate variables"). He manages to formulate a relationship between the uncertainties with which We can measure these conjugate variables. If I call the variables **"A"** and **"B"** (for example "position" and "momentum"), and if I call **"ΔA"** and **"ΔB"** their respective uncertainties, then we have that ΔA multiplied by ΔB will always be greater than or equal to (guess what?) the Plank constant **"h"**. For example, position and speed (actually "momentum") are two conjugate variables. I'll try to show you this with the example below (fig 5.6):

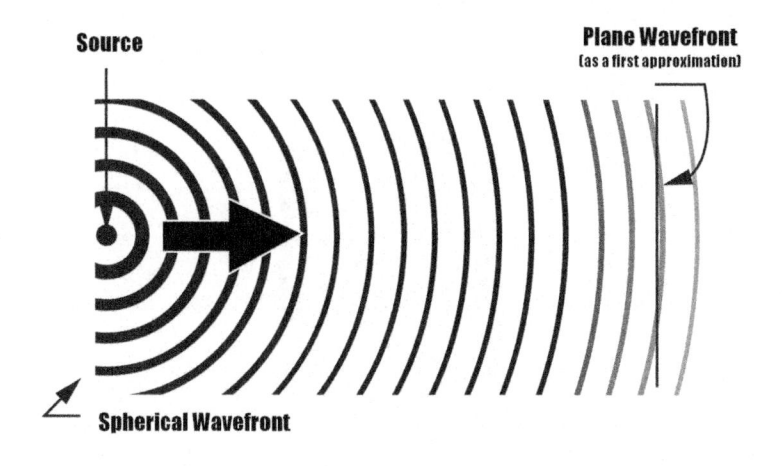

Fig. 5.6

The drawing above represents the wave associated with an object that is shot horizontally towards the right from a source. Positioning ourselves a little away from the source, the wave has meanwhile become a "plane wave", so defined because the crests resemble more straight lines than arcs of circumference. I remind you that, regardless of the speed "v" of the object, the wave travels in the same direction but with the speed of light "c". Why did I make this clarification? I did it in order to reiterate that almost immediately the wave tells me that the position of the object, far from the source, and therefore now free from the effects of any perturbation, is undefined, along both axis: let's focus along the vertical axis, it is clear we do not know anything, since the crests are vertical lines, therefore each point of the vertical axis has the same probability; the position is undefined also along the X axis, although we see points of maximum and minimum (but I won't bother you with this part of the story).

We think we are smart and can play a trick on Heisenberg, precisely on the vertical position. We know that our object's vertical speed is zero: indeed, the wave is traveling only along the x axis, and is shows a constant value at any point along the vertical position; we place a slit (only one this time) so that we expect the vertical position will be at least inside the segment subtended by it. Then we can make the slit as small as we like, thus finding a way to make the wave determine for us both the vertical speed (which we knew is zero and should have remained so since no one has touched the particle) and vertical position (thanks to the slit) with certainty. However, look (fig. 5.7) at what happens to any wave when it passes through a slit, comparing the scenarios between slits of two different sizes (again we are seeing real water waves).

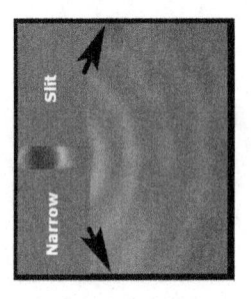

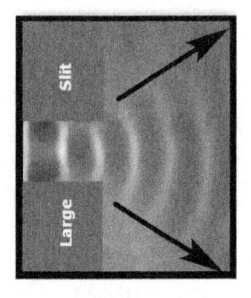

Fig. 5.7

The more I squeeze the slit the more the wave widens, thus the higher is the contribution of waves directions no longer just perpendicular to the slit which means we have introduced an uncertainty on the speed along the parallel direction (even if I don't touch the particle represented by that wave).

I hope with this example you can see that the more I tighten the uncertainty on a variable (in this case by reducing the size of the slit), the more the uncertainty on its conjugate variable (the speed parallel to the slit) increases.

Now, let's move to Schrödinger's Cat Paradox.

As stated, Schrödinger was one of the skeptics. In order to criticize everything I have told you so far, i.e. the particle that passes through both slits if I don't measure it, he enunciates a paradox, which I am now going to modify so I can reuse the example of the two slits.

I put a cat in a box; I add a gun that shoots an electron that must pass through an obstacle with two slits; immediately beyond one of the two slits, I put a container of poison that is broken if crossed by the electron, so that the cat dies. Obviously, the cat is alive if the electron passes through the other slit. I close the box, therefore I don't know where the electron passes through, so my cat is both alive and dead; until I open the box (fig 5.8).

Fig. 5.8

With this statement, Schrödinger wanted to emphasize the absurdity of the Copenhagen interpretation.
Do you agree with him? Wrong.
I hope someone has noticed: in reality, it is the container with the poison that performs the measurement indicating which slit the electron passes through. I don't need to open the box. Even if I don't open the box, quantum mechanics tells me that there is no overlap (or interference) between the two possibilities. So, the cat is alive or dead before I open the box, even according to quantum mechanics! Okay, that's enough. We could keep going for another thousand pages, but you must be tired. The next chapter, on the other hand, will help you experience the first great success of this theory, namely the physics of atoms.

Chapter 6. Atoms – Quantum Mechanics' Cup of Tea

We now come to experience what was the first great success of quantum mechanics: the physics of atoms.
In fact, we will see how, according to classical physics, the universe (and also ourselves) cannot exist; then I hope I can show you how only quantum mechanics at the moment is able to "fix the problem". Spoiler alert: it won't be so straightforward, but I hope you'll see some light at the end of the section.

6.1 – Atom's Lifetime

I imagine you are all well aware that science has concluded that we are all made of atoms and, moreover, that the atom has more or less the structure shown in the figure 6.1 below.

Fig. 6.1

The story of how we got here is fascinating, and could be the subject of a specific book, but for now we just need to remember that classical physics arrives at that structure around 1911. Indeed, you must know that atoms cannot be seen, no one has ever seen them and no one will ever see them. However, with the same trick used by Davisson and Germer when firing small bullets at a target and seeing how they scatter off, New Zealand physicist (and chemist) Ernest Rutherford convinced everyone that the internal configuration of any atom was the one shown in the above figure. In this case, the small bullets shot at the target were the alpha particles, discovered a few years earlier.

A small clarification: the figure is not to scale, as that same experiment had estimated that the size of the diameter of the nucleus (the ball you see in the center) is ten thousand (10,000 !) times smaller than the size of the diameter of the atom.

Now let's recall what I wrote in chapter 2.

I am referring to Maxwell's equations that predicted, among other things, that a charged body that moves with non-uniform motion emits waves, called electromagnetic waves.

The electrons (the balls in the outer circles) are charged bodies and, in that figure, they are certainly not moving in a straight line, that is, in a uniform motion. Thus, electrons must also emit waves! But if they emit waves, then they must gradually lose energy (the one they give to the wave); if they lose energy then they slow down; if they slow down, sooner or later, they will fall onto the nucleus that is attracting them (nucleus and electron have opposite electrical charge).

How long does it take before 'this disaster' takes place after – so to speak - an atom is created, according to classical physics? The answer is: half of a tenth of a billionth of a second, so not exactly the several billion years which is the estimated age of the universe, nor even our age. Is there a way out of this problem?

6.2 – Bohr's Atomic Model

The first attempt comes from Danish physicist Niels Bohr, around 1913. You have already heard of him when I told you about the Copenhagen interpretation in the previous chapter.

Out of the blue, he makes up a purely empirical rule, i.e. a rule which does not derive from the equations of existing theories. In a nutshell, his rule prohibits electrons from emitting waves, except in some particular and rare cases.
What does this rule say?
We need a premise: you surely have realized that physics needs numerical quantities to describe Nature's behavior. For example, in order to describe a body of mass "M" traveling at a certain speed "V", it uses a quantity called "momentum", that is the product of M times V. It does so because, by using this quantity, the equations of motion of bodies are easier to solve. But also, let's face it, this formula makes practical sense: the product of mass times velocity can be useful in some circumstances, like for quantifying the different damage I expect to suffer if I get hit by a fast ping pong ball rather than by a slow train.
For rotational motions, like motions on closed orbits, the "twin quantity" of the momentum is called "angular momentum", which is a mathematical expression of mass, speed and distance of the mass from the center of the orbit. I'd like to draw your attention on the units of measurement of this "angular momentum": from its very definition, they are Kilos (or grams or pounds), multiplied by kilometers per hour (or other unit of measurement of speed), multiplied by meters (or other unit of measurement of space). Well, believe it or not, these are the same units of measurement (in jargon, "dimensions") of Planck's constant "h" (here it comes again!!). End of the premise.
So what does Bohr do? He claims that, since for electromagnetic waves - remember that in 1913 we still do not have De Broglie's wave - the ratio between energy and frequency must be an integer multiple of "h", as stated by Planck and Einstein with the expression $E = hf$, then perhaps also the angular momentum of the electrons in the atom must be an integer multiple of h, starting from value 1h.
There is no logic behind his statement, he is only proceeding by similarity.
However, what do you get out of it?

With this additional constraint, it results that the poor electrons can have only some values of angular momentum – translated, some values of rotation speed. As a consequence, it is impossible for the electron in the atom to slow down gradually, because if it starts from an allowed speed value, it has to go through those that are not allowed before moving on to the next allowed value, and therefore the process cannot take place (see drawing 6.2).

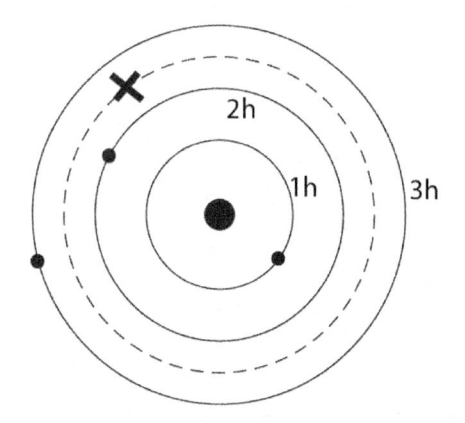

Fig. 6.2

Following this rule, the electrons keep happily rotating in the orbits that are allowed by this quantization of the angular momentum (in the drawing the orbits h, 2h and 3h, etc.); moreover, they do not emit any waves, because they cannot slow down since they can never be on an orbit like the dotted one. This orbit instead would be allowed by classical physics, because with the equations of classical physics there is always a speed that allows an electron to rotate in any chosen orbit.

I must confess that Bohr focuses on the hydrogen atom, the simplest to study, because it has only one proton and one electron. However, this idea, philosophically speaking, prevents the collapse of any atom.

Anyway, as I have told you to the point of boredom, any new idea must be associated with quantitative predictions, which some experiment must then confirm.

The first experimental confirmation of Bohr's hypothesis is in the next paragraph.

6.3 - Spectroscopy

You have already met spectroscopy in this book.

It happened in chapter 3, when we introduced the Plank constant and when I told you that physicists studied the spectrum of colors resulting from the light emitted by objects at various temperatures.

Well, in the case of incandescent objects (for example the old filament bulbs), I told you that the spectrum is always a full rainbow, with the presence of all colors, even if their incidence varies according to the temperature.

However, already by the early 1800s, scientists had realized that by causing some kind of electric spark between two ends of a thin tube, possibly filled with a single gas, this tube would light up (today we call it a neon lamp regardless of the kind of gas that is inside). Most importantly, they realized that even in cases where the light looks white (such as with hydrogen lamps) the resulting spectrum is not a rainbow, and instead only some colors show up. For example, the colors of the hydrogen lamp are 4: red, cyan, blue and violet, while neither yellow nor green appear.

In order to better understand what I mean, I recommend that you look at this VIDEO[3] (more or less 6 minutes).

No one had raised the problem of finding the origin of this phenomenon, while everyone focused on its usefulness. Indeed, it was clear that each gas emitted a particular type of spectrum, as if it were a signature. Thus, once all the spectra had been cataloged, one could trace the types of gas of an object that shines - that is, of a star – just by studying its spectrum.

Bohr's hypothesis - with a few small tweaks – actually predicts this phenomenon.

All we need to do is to assume, once again without any reasoning behind it, that, although electrons cannot gradually slow down, they are allowed to "suddenly" jump between two permitted orbits. This, obviously, without ever being allowed to go below the lowest orbit, that of the value "1h", and provided that the target orbit is not already occupied.

You probably have become very impatient by this sequence of ad-hoc assumptions, which even involve contradictions (why can't I slow down but I can make the leap? And then what does it mean to make the leap suddenly?).

Unfortunately for you, however, using all these hypotheses and running the calculations to predict the energy differences between the levels h, 2h, and so on, initially for the hydrogen atom and then also for the others, we obtain exactly the energies associated with the 4 experimentally observed colors.

Using Bohr's theory, which, I repeat, was initially created only to justify the stability of atoms, we could predict how many and which colors were part of the spectrum of various gases. Clearly, the electric spark must somehow cause some electrons to jump (again suddenly) forward and always only towards one of the permitted orbits with a higher h value than the current one. Then, although we do not know why (actually we do but it is a long story), the electron wants to "go back" to one of the lower empty allowed orbits, while emitting a radiation of energy equal to the backward jump. There would still be one assumption to verify directly: does the electric shock, or any external energy source, really manage to supply only some specific amounts of energy and not just any value?

This was already confirmed by the absorption spectrum of gases (I spare you this part), but the direct confirmation of this assumption comes in the experiment conducted by Franck and Hertz (Gustav, not Heinrich) in 1914.

I suggest you watch this VIDEO[4], which I really like and that, even if very old, I recommend to everyone. This time it's almost 29 minutes long, but it's really instructive for a lot of things.
For those who do not want to see the video, obviously you can find a lot of literature on the internet, including Wikipedia. However, in a nutshell, instead of producing an uncontrolled shock, Franck and Hertz continuously send bullets (electrons) with known speed (and therefore energy) against a sample of hydrogen gas inside a container of glass and count how many electrons, in a certain time lapse, emerge from the other end (see drawing 6.3).

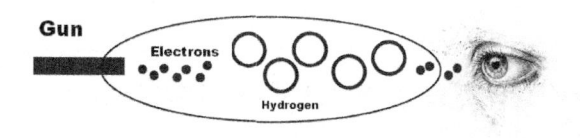

Fig. 6.3

If my bullet - an electron - did not give energy to any of the hydrogen atom's internal electrons after colliding with it, then it would bounce with practically the same speed of incidence. In fact, the atom weighs two thousand (2,000) times more than the bullet: it is like a ping-pong ball bouncing off a basketball, moreover without giving up even a gram of its energy. So, IF all this takes place, the bullet will keep bouncing from another atom, and so on. In short, there will be a more or less constant fraction of the incident bullets that get to where I drew the eye, after many bounces. That's exactly what happens, except for some precise velocities of the incident bullets (remember that in this experiment we have the possibility to determine a priori what this velocity is). At these particular speeds, the fraction of emerging bullets collapses dramatically. This means that they have transferred practically all their energy to the hydrogen atom, or rather to one of the electrons of the atom, which in turns can make the leap forward onto one of the allowed orbits, as predicted by Bohr. Needless to say, this phenomenon is observed precisely when the incident bullets have exactly the energies predicted by the model. Furthermore, photons are also observed, as a sign that the electron of the atom "goes back".

However, the theoretical basis (so to speak) that manages to predict all this phenomenology is Bohr's theory of the quantization of the angular momentum (which means the energies, i.e. the orbits) of the electron in the atom. However, I repeat, this theory is just an artfully advanced ad-hoc theory.

In the next paragraph, we will instead try to see how true quantum mechanics, - that is the one with the De Broglie's wave and all the rest- contains "Bohr theory" for FREE among the solutions of its equations; a bit like Maxwell's equations contained in themselves the existence of electromagnetic waves.

6.4 – From Orbit to Orbitals.

So let's give it a try.
How do the equations of quantum mechanics solve all problems?
We obviously will not write even a single formula, and this will limit our capabilities. I mean, it is true that all solutions magically come

out only after writing and solving the equations with which quantum mechanics describes Nature. However, we will try instead with a pictorial trick to see how the imperative of having to describe even material objects by means of a wave, forces us to quantize closed orbits, just as Bohr claimed. Hopefully we can be satisfied after achieving this result, keeping in mind that this quantization also solves other problems related to atomic phenomena, like for example spectroscopy, as we saw in the previous paragraph.

In practical terms, we will see how quantum mechanics makes us switch from the concept of orbit to that of what is called "orbital". Perhaps many of you have already heard of atomic orbitals and their particular "shape". (See figure 6.4 for some samples).

S-type Atomic Orbital

P-type Atomic Orbitals (3)

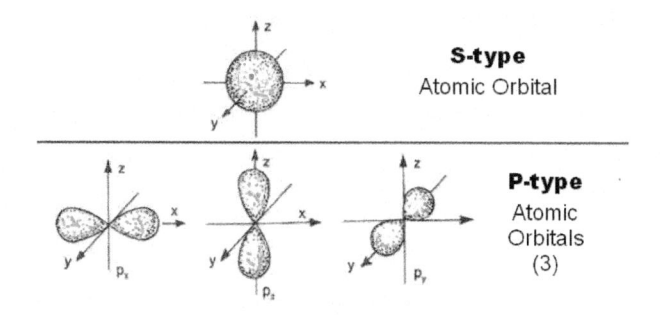

Fig. 6.4

Surely you have heard that it is the equations of quantum mechanics that generate these shapes, and that is true, and you will be surprised to see that only one case is a sphere.

So we will ALSO see how the sphere and the other strange shape come out (there are many others but we stop here) from quantum mechanics.

Again, we need a premise, or rather a clarification. Let's start with quantum mechanics and what makes it different from classical mechanics. In classical mechanics, the main equation is Newton's equation, known as the second law of dynamics: F = ma, which states that if we apply a force of intensity **"F"** to a body of mass **"m"** it will undergo an acceleration of value **"a"** such as to satisfy the fact that "m" multiplied by "a" equals the value "F". With this formula, once we know how to represent the forces through arithmetic expressions and once we know the initial speed of our system, we are able to predict what happens to it at any future moment in time. The power of this formula has allowed us to predict the motion of the planets, to discover new ones, to design bridges, roads and houses. It has also allowed us to understand how gasses and fluids behave once we were able to represent them as a set of small objects (which was Newton's attempt ALSO with light), in short, to predict (almost) everything. Quantum mechanics goes one step further. It Transforms Newton's equation into another equation, known as Schrödinger's equation (already heard this name?). This new equation, which I am not going to write down because it would not tell you anything, allows us to understand what happens to the wave associated with an object of mass "m" with initial velocity "v" when the object is under the influence of force "F".

The true expression does not use force but a quantity that in jargon is called "potential", and which is closely linked to force. So, while in classical mechanics we take an object and apply Newton's equation directly to it, in quantum mechanics we initially must associate the wave with the object, using De Broglie's rules, then apply Schrödinger's equation to see how the wave evolves; from the wave we revert back to the new values of the "classic" properties of the object (and not even all of them: remember Heisenberg's uncertainty principle).

I wrote this long premise just to try to draw your attention to the importance of the wave.

Let's try to visualize with a graphic trick what properties this wave must have when it describes an object on a closed orbit.
I will try to convince you that if you choose a hypothetical orbit, whose circumference has a definite length, you can't use just any wave to describe the object running on it steadily. By steadily I mean that it keeps spinning round, as the Moon does around the Earth, the Earth around the Sun, or an electron around the nucleus.
I want to convince you that in order to describe this phenomenon, the wave must be such that its "wavelength" is an ENTIRE submultiple of the length of the circumference of the orbit.

To see what I mean, let's look at the sequence of drawings 6.5.a below:

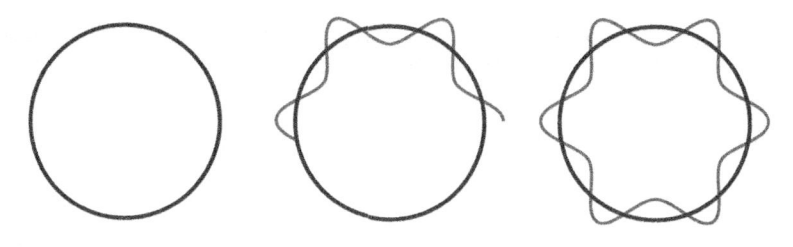

Fig. 6.5a

On the left we have a circumference, in the center we begin to draw a wave on it, which however is not just any wave: indeed, on the right, we realize that this wave closes perfectly on itself, which means that the circumference is an integer multiple of the wavelength.
Instead, let's see in the drawings 6.5.b what we get if I really pick a random wave.

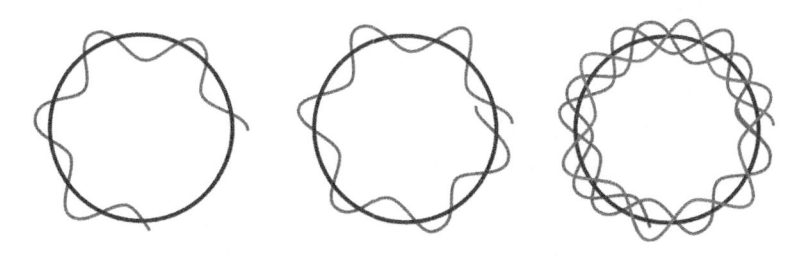

Fig. 6.5b

As you can see from the central drawing, the wave does not close in on itself and therefore if I "continue" to move along the orbit I will always get waves in different positions, as I hope the drawing on the right shows. Then the combined effect of all these turns will be that the wave cancels itself out and the overall wave will be zero, whereas in the previous case the wave remains "stable" (it is called a standing wave).
Only the "stable" wave will describe a body that turns "steadily" on a circumference, while the "random" wave describes nothing.
It is therefore imperative that the circumference has a length that is an integer multiple of the wavelength, or, vice versa, that the wavelength is an integer submultiple of the circumference.
That said, let's recall De Broglie's rule. The wavelength is linked to the object's speed, and even classical Physics tells us that an object must have one and only one precise speed in order to run along a

certain circumference when the binding force is fixed. So what? Well, you do the math!! Take a circumference, use classical physics to calculate the necessary speed: if the wavelength associated to that speed is NOT an integer submultiple of that circumference, then that orbit is an impossible orbit, in other words, prohibited. Only those circumferences whose associated rotation speeds give rise to a wave whose wavelength is an integer submultiple of the circumferences themselves survive. Thus, we reached at this conclusion using a graphic "trick" because the rigorous procedure requires that we solve Schrödinger's equation which deals with the binding force between the nucleus and the electron. Hopefully we can now glimpse how the mathematics of waves applied to closed orbits offer us - for FREE - those hypotheses of forbidden and allowed orbits that Bohr had instead pulled out of the blue. Therefore, we have seen how quantum mechanics automatically generates Bohr's hypotheses, which proved to be effective in describing even other properties of atomic phenomenology.

We can take one more step. Let's see, once again not with a rigorous procedure, how both spherical and the strange shape orbitals come out. The "orbitals" are the shape of those waves, which, I recall, quantify, in each point of space, the probability of finding an electron at that point, at a certain instant. So let's look at the drawings 6.6 a little far down.

In the upper portion of the drawing, I opened up a circumference (solid line) and transformed it into a straight line segment. On the upper left, I added a wave (dotted line) whose wavelength coincides precisely with that of the circumference, therefore multiple 1; on the upper right instead the wavelength is half the circumference, therefore mutiple2. I haven't even tried to draw the higher multiple cases. In the center part of my picture, I tried to draw what we get when we start folding the solid line to reconstitute the circumference that I had previously straightened, and finally, at the bottom edge, the final result. If we observe the dotted lines, and maybe try to picture them in 3 dimensions, the bottom left reminds us of a sphere, while the bottom right the other structure with two lobes (I hope that the position of the areas marked with the + and − sign with respect to the continuous line do help somehow).

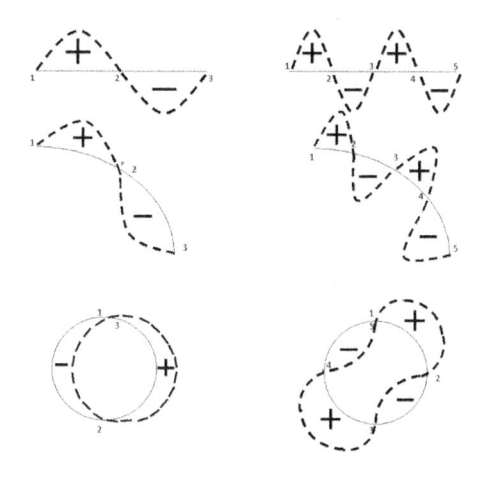

Fig. 6.6

6.5 – So What?

I hope I convinced you that quantum mechanics, born to explain the photoelectric effect, then developed with the extension of the particle/wave duality to matter, carries the solution to all the phenomena concerning atoms within itself.

We have only seen two of such phenomena, but believe me when I tell you that the shapes of those orbitals explain the chemical properties of the atoms they refer to. It is true that many of these properties had already been identified and perfectly cataloged by the Russian chemist Mendeleev in 1865, with his famous periodic table of the elements. However, even in this case it was an ad-hoc exercise, although useful and powerful. Even this now becomes just a particular case of a broader theory. After the atoms, all the phenomenology concerning the nuclear world, then the sub-nuclear world, all the way down to what is now thought to be the fundamental ingredients of matter and forces, needs quantum mechanics to be described.

Even the classical equations of motion, for a bowling ball for example, could be replaced with those of quantum mechanics, but, in this case, the results would be exactly the same as the equations of classical physics; it therefore makes no sense to get fancy for nothing and we continue to use classical physics. Quantum mechanics must be used starting from atoms and gradually descending into the infinitely small, even if, as you may have heard, we are now trying to exploit its bizarre features to manage some macroscopic phenomena, such as the quantum computer or quantum cryptography, to which I will provide a very short explanation in the next chapter.

Chapter 7. Practical Applications

As I have already told you, in practice without quantum mechanics we would not even be able to predict our own existence, let alone all that follows.

However, I know that you want something less ethereal, thus I could decide to tell you about those topics that are very cool today, i.e. quantum computer, quantum cryptography or, for the more accustomed, quantum entanglement (which is underlying quantum cryptography).

Okay, in order not to disappoint you, I will do it in the last two paragraphs of this chapter, only after having addressed some other topics that I think are more significant to fully understand the "practical applications" of Quantum Physics.

As a start, I will simply mention that all the medical diagnostics applications, such as CAT, PET and MRI, or some therapies such as Hadronic Therapy, exploit the intimate properties of atoms, if not of nuclei. Properties that we could have never even envisaged without quantum mechanics. Even ordinary radiographs, while using X-rays that were discovered well before quantum mechanics, benefit from the fact that with the latter, we can predict how radioactive materials behave; an impossible result with classical physics, since even radioactivity is a nuclear phenomenon.

Anyway, before getting to the cool stuff, in the next four paragraphs, I want to expand on an application that influences the life of all of us at every second: i.e. electronics, actually, solid state electronics.

Yes, without quantum mechanics we could not have imagined creating tablets, mobile phones and PCs, but nowadays even any household appliances and cool electronic stuff you find everywhere.

Indeed, it is only thanks to quantum mechanics that humanity could imagine building an object that we call a "semiconductor diode", followed by the transistors, microprocessors and so on.

I will tell you how we came to think that a semiconductor diode can exist and how this can replace an object called a thermionic valve (the first diode), with the result of making electronic circuits smaller, faster and more practical, in short, such as to allow us to create all the gadgets I mentioned earlier.

7.1 – Electronic Devices

What is the difference between an electrical circuit and an electronic circuit? Both use electricity, but the first one is "limited" to bringing energy from a source - a battery - to a user, such as a light bulb or an electric motor. The second one, on the other hand, is able to compare scenarios, make calculations and make decisions. In reality, it always and only calculates sums (or differences, that are always sums. I won't go into detail, but I will just remind you that multiplication is a sequence of sums, division is a sequence of differences, and so on). But what about scenarios comparison? Well, if I can represent each scenario with a number, I can make the comparison by looking at the difference between the two "numbers". Even making a decision can be represented with a sequence of numbers. Indeed, every decision can be seen as a tree of elementary decisions, that is, either I choose A or I choose B. In the example below (figure 7.1) I hope I can graphically show how, with two elementary choices, I can reproduce a choice between three scenarios, A, C and D.

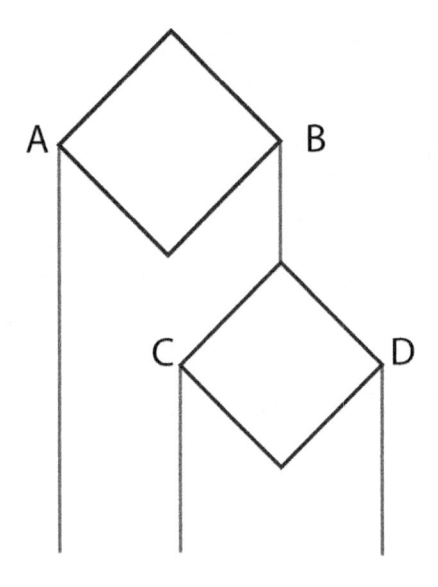

Fig. 7.1

We can read figure above in the following way: if I choose A then it is A, if I don't choose A (case B), then I will have to choose between C and D, and this way I can represent the three scenarios using only the symbols 1 and 0 (A = 001, C = 010, D = 100).

Having ascertained that we can reduce everything to numbers and sums, let's focus on sums.
We know how to add up with decimal numbers, but there is a subtle way to add up and that is by using binary numbers.
What are decimal and binary numbers?
Let's start with the decimal numbers, the ones we normally use. We have 10 symbols, from 0 to 9, with which we represent quantities. If we have a quantity that exceeds 9 by 1 we write 0 again, but we put a 1 on the left, 10, and so on.
We can do the same with binary numbers, where the symbols are only two: 0 and 1. In this case, when the quantity exceeds the value "1" by 1 we put a 0 and to its left we put a 1, that is we write 10, which in binary is the equivalent of our 2, just as 11 equals 3, 100 equals 4, etc.
Now, there is no practical advantage for us human beings to use this approach: the numbers would become very long to write and to read when there are so many characters. The advantage is that, if one uses this binary system, even the numbers and operations, as well as

the decisions and comparisons I described earlier, eventually become just a sequence of 1's and 0's.

I hope I gave you the feeling that by finding a clever way to manage numbers that contain only sequences of zeros and ones, we can represent any form of reasoning.

So, if I find a way to ensure that an electrical circuit can generate ones and zeroes, I will be on track: it will just (so to speak) be a matter to properly organize these circuits together.

Well, that's exactly what these electronic circuits do: process large sequences of 1s or 0s; in paragraph 7.3 we will see in part how this is possible. In the meantime, let's start to see how to generate those 1s or 0s.

7.2 – The Thermionic Valve

Around the early 1900s, the Thermionic Valve was developed, which indeed performs just what is needed in our case: generation of 1s or 0s. How does it work?

I am not saying anything original; on the contrary, I will be rather quick and dirty.

If we take a thin metal wire and run some current through it, we realize that after some time, it will become bright; by inserting it into a closed glass container - inside which we created some vacuum - we have invented the light bulb. It was also understood that wrapped around this overheated wire, indeed precisely because of this overheating, one would get a cloud of free electrons.

What happens if, in addition to the wire, I also put two other metal objects inside that same glass container and I connect them to a separate battery? Well, if one of the two objects is connected to the negative pole of the battery and the other one to the positive pole, an electric force is created in the space between these two objects that would tend to push electrons away from the negative and towards the positive. If the object connected to the negative pole is closest to the heated wire, which in turn is surrounded by a cloud of electrons, some electrons waving will end up between the two metal pieces, and PRESTO! they will be pushed towards the positive pole. Basically, we created a current between the negative and the positive pole. If, on the other hand, I invert the battery, the electric force between the two metal objects does the opposite: it tends to push any electrons from the distant object (now connected to the negative pole of the battery) to the one close to the heated wire. However, there are no electrons to push from the distant object in this scenario, since they are already close to the heated wire and therefore they remain there, and no current is created between the two objects connected to the battery.

Normally, with any piece of an electrical circuit, if I place the battery in one direction, the current runs in one direction, while if I invert the battery, the current runs in the opposite direction. Instead, in this case, I have an object that, if I connect the battery in one direction, I have current, while if I change the verse of the battery polarity I don't: in other words, in one case I have a 1, in the other case I have a 0.

It is precisely with these valves that up to the 1960s the first electronic objects were built: ranging from computers, which occupied entire rooms, consumed tons of energy and were very slow compared to the current ones, to the first television sets. Someone my age or older will surely remember those TVs.

I think it is obvious that with those "light bulbs" (fig 7.2) we could forget about smartphones and so on!

Fig. 7.2

7.3 – Off the Record – Two Electronic Circuits

For the curious readers, I will try to show two basic electronic circuits, which are called "logic gates": the AND gate and the OR gate.

The AND gate analyzes two input signals: if both inputs are equal to 1, it returns the value 1, otherwise it returns the value 0. The OR gate does the exact opposite: if both inputs are equal to 0, it returns 0, otherwise it returns 1. Trust me when I tell you that any algorithm and any operation can be represented by an appropriate mixture of these ANDs and ORs, plus a third gate, which is called NOT, and does nothing but analyze a single signal, returning 0 if the input is 1 and vice versa.

So, let's see the OR gate (figure 7.3):

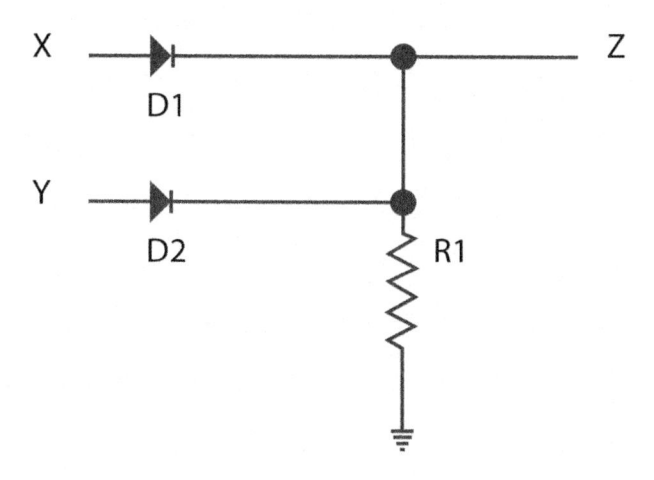

Fig. 7.3

X and Y are the two input signal, Z is the output we measure. I can claim that each of the two is 0 if I don't connect it to an external battery or 1 if I connect it to a battery. If we connect such batteries, their negative poles share the same "ground" with resistance R1.

The symbol ▶︎▎ represents the diode itself and when it is drawn in this way it means that it is positioned so that the current can only flow from left to right. Let's see what we get:

If I connect the battery to X, then I say that X is equal to 1. The battery wants to run a current from X to the ground; it has no obstacles since the diode D1 allows the passage of current from X to the right. The current, after passing through the resistance R1, comes straight to the ground. I point out that the current cannot go towards Y because the diode D2 prevents the current from passing, despite at this moment also Y behaves like the ground because no battery is connected. So, the current only flows through the resistance R1. Everything therefore is working as if we had an electrical circuit made only by resistor R1 and battery connected to Z, so we can say that Z is worth 1.

Same thing will result if I connect the battery to Y instead of X: in fact, the current flows from Y to the right, it cannot go towards X (although X is grounded) because the diode D1 prevents the current flow, therefore the only outlet to ground is R1: we are again as in the previous case.

Obviously, if both X and Y are 0, nothing flows, and if both are 1, again their current is vented only onto R1. So Z is always worth 1, except when both X and Y are 0.

Let's now look at the AND gate (figure 7.4):

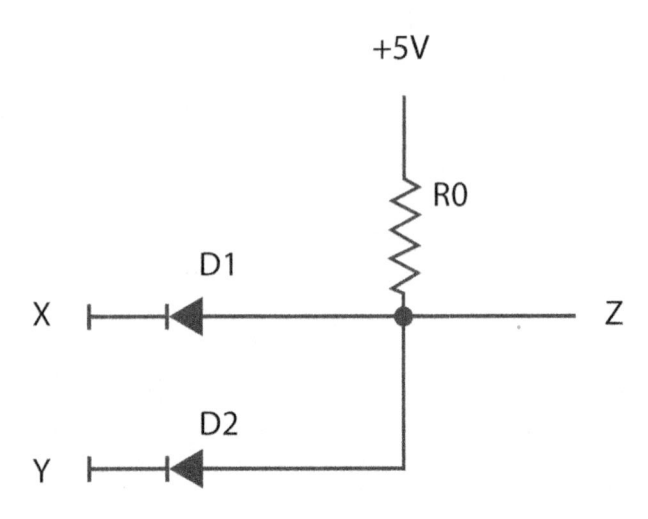

Fig. 7.4

Here we already have a battery that wants to run a current, make it flow through the resistance R0 and then go to the ground. The current has three possible paths: towards X, towards Y and towards Z. If X is 0, that is, if it is grounded, the current can go towards X because the diode D1, this time positioned so as to allow the flow from right to left, allows this passage of current; while if X is 1, that is, it is now connected to another battery, the current cannot go towards X. And the same goes for Y. Basically, the current that starts from above R0 will go all towards Z only when both X and Y are worth 1, otherwise it will disperse between two or all three paths. In other words, there is only one scenario that is equivalent to having only the 5V battery, the resistance R0 and the output Z: the one in which both X and Y are worth 1.

I hope I have not confused your ideas too much, but I just meant to give you the feeling that "yes, maybe with these crunches you can do what you say".

7.4 - Semiconductors and Diodes

In the gates I drew earlier, the role of diodes was played by thermionic tubes, so you can imagine the size of something that must combine hundreds or thousands or more.

But thanks to quantum mechanics it has been possible to envisage (starting from around 1928) and to actually make (around 1940) objects that behave like valves, but are much smaller: junction diodes.

What was the idea?

I will try to show you that, thanks to the concept of atomic orbitals, we managed to assume that it is possible to create objects with certain materials where we have a "geographical" and stable contrast of opposite electrical charges, in jargon a "junction". I will also demonstrate that such a junction behaves like a thermionic valve.

Let's start from the junction and its properties, later on we will see the importance of quantum mechanics specifically with its allowed orbits.

In the drawing 7.5 you see that there is an array of positive charges that is in front of an array of negative charges.

```
              |  -  |  +  |
              |  -  |  +  |
              |  -  |  +  |
              |  -  |  +  |
              |  -  |  +  |
```

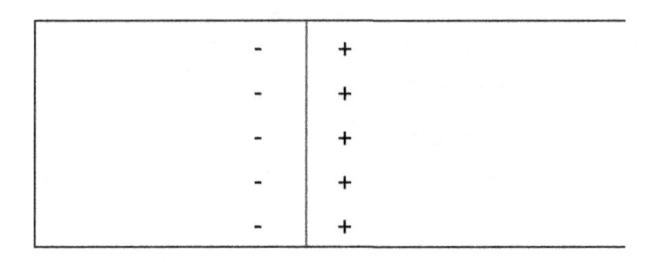

Fig. 7.5

We get this junction because, with specific materials, it is possible to "drug" the right half in such a way that, while remaining electrically neutral, more negative charges will be free to move than positive ones, and on the left half hand side more positive charges will be free to move than negative ones (I know, for now drawing 7.5 looks like the exact opposite). Ok, for now let's assume that I can get this different concentration of free-to-move charges between the two halves: the free charges will each tend to drift into the other half, in order to equalize the concentrations, precisely like two liquids with different salt concentrations. Some pairs of opposite charges, collapsing with each other during the journey, will recombine, effectively canceling each other out, but some will manage to cross over. Up to where? How many? There will be positive charges that, having managed to cross over from the left half, will now result in excess in the right half, while there will be negative charges in excess in the left half (this time exactly like in the drawing 7.5). Those in-excess opposite charges will tend to attract each other, until this emerging force of attraction, which would have the tendency to bring the positive charges back to the left side (where there is now the excess of the negative ones) and the negative ones back to the right side, will compensate the drifting original tendency. This creates a situation of equilibrium between these two opposing trends, the result of which is the double array of charges highlighted in the drawing.

Now let's see if this junction behaves like a valve. If I can somehow connect the negative pole of an electric battery to the right side and the positive to the left one, what do I expect to obtain?

The negative pole tends to push any free-to-move negative charges (the very ones with which I initially drugged the right side) to the left, while the positive pole tends to push any positive free charges (precisely those with which I initially drugged the left half) to the right. In essence, the battery placed in this way increases the drifting tendency that the free-to-move charges had initially. In other words, it breaks the equilibrium and favors the continuous drift, which indeed is like a current that flows from the positive to the negative pole of the battery. In short, I obtain exactly what I expected if I had connected a conductor in that very same way to the same battery. As a result, we have a current that runs in that direction.

Vice versa, if I connect the positive on the right side and the negative on the left one, I end up helping the force that opposes the drift, therefore everything remains stationary and I have no current.
So, once again, I created a diode, where the current passes only if I connect the battery in one way and not the other. But this time everything happens in much smaller dimensions than the valve and without the need to have a continuously overheated wire. There is therefore a lower expenditure of energy and the process runs more quickly, given the spatial scale of these phenomena.
What does quantum mechanics have to do with it?
Please remember that I wrote that I can make this junction because I can have two areas that I can "drug" with excess free-to-move charges. This is possible thanks to the concept of "semiconductor", which is a direct consequence of the fact that not all orbits are allowed in atoms. Therefore, without quantum mechanics we would never have guessed these results.
Let's see how.
In the previous chapter, we stated that quantum mechanics offers us the complimentary conclusion that in atoms only some and not all orbits are allowed.
Note: since each orbit corresponds to an energy, from now on I will use the term "energy levels" instead of "orbits".
You must also accept that quantum mechanics also states that in any given energy level there cannot be more than two electrons. This step is just as important as that of the forbidden orbits. As a conclusion, it is clear that the electrons will occupy different levels instead of staying all in the same one. It is precisely the combination of these two rules that leads the elements to have chemical properties that can be represented with the periodic table of the elements (which, I recall, was almost completed over half a century earlier thanks to empirical observations and to the cataloging skills of the Russian chemist Mendeleev).
But let's move on: when we move from dealing with a single atom, or atoms, in a gas, which can be considered made of many single atoms, to dealing with atoms in a solid, we are in a condition where

atoms are wedged together and influence each other. The result is like if energy levels were being pooled from among a number of neighboring atoms, as shown in the drawing 7.6 below:

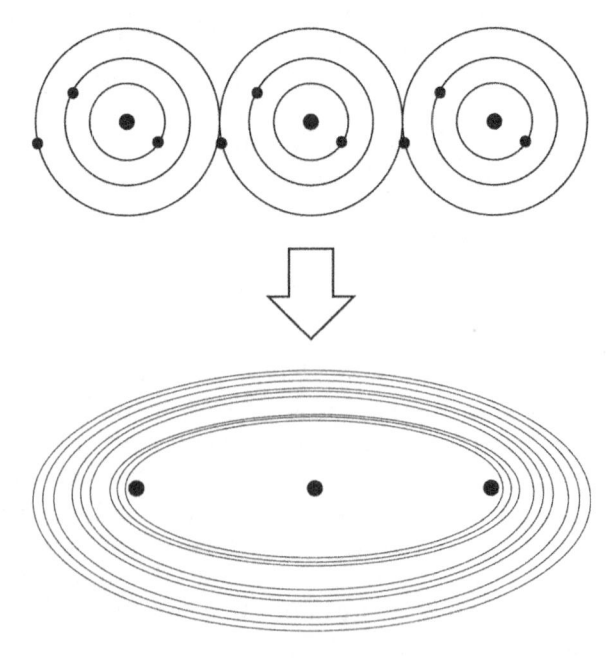

Fig. 7.6

Each atom taken individually had, let's say, just three energy levels, as in the upper part of the drawing; if the three atoms, being very close to each other, share their energy levels, their electrons behave as if they were affected by this proximity, resulting into three groups of only three levels.
But let's increase the number of atoms that affect each other. For example, consider that in 2 grams of silicon there are about ten thousand billion billion atoms: I hope it is reasonable to assume that they will create energy bands. But careful, quantum mechanical equations can predict how many bands, how wide they are and how far they are from each other, to some extent. Ok, each atom contributes with its electrons and all these electrons will find their place in the various bands, starting from the lowest.
Based on what we just said, we can have three scenarios:

1 - Almost all electrons occupy a certain number of internal bands and only a few occupy the next outermost band, no matter which one; most important is that it is the outermost band. These electrons are the least bound to their nuclei, moreover they are few in a band made up of many possible levels. Thus, they are free to move. Cases like these coincide with those elements that result being "conductors" of electric current.

2 - Conversely, I could have cases in which the outermost band is almost completely, or totally, occupied by electrons, while the next band is energetically distant. In this case, the electrons are almost imprisoned; this condition indeed coincides with the elements that we call "insulators".

3 - The calculations foresee cases similar to insulators, with the energy bands full of electrons, but where the distance to the next band is almost zero. In these cases, for example in silicon, it is not impossible that some electrons jump into the next band, thanks to thermal energy, even at room temperature. These elements behave halfway between insulators and conductors and we call them "semiconductors".

Do you remember that a few paragraphs ago, when I spoke of the "junction", I referred to free-moving charges, both positive and negative? Well this happens in semiconductors. The negative ones are the electrons jumping in the outermost band. And the positive ones? We call them "holes", and they are the voids that are left in the band below by the jumping electron. Indeed, when an electron jumps into the outermost band and is no longer bound to any atom, that atom has now one fewer electron, thus taking on a positive charge. But nothing excludes that one of the electrons of the other neighboring atoms goes to fill that hole making this other atom positive; then another one fills in and so on. In short, this hole of negative charge, which therefore has the same properties as a positive charge, moves almost freely. In this way, I hope I told you how quantum mechanics foresees that certain materials exist, which we call semiconductors and which are the materials where these free charges of opposite sign coexist.

One last step is missing.

In fact, those of you who didn't get lost will have noticed that the holes are created when an electron jumps, i.e. in the end the number of free positive and negative charges is identical. However, when I talked about the junction I said that there was an area with more free negative charges and one with more free positive charges.

How can this be?

Long story short: if I manage to contaminate that piece of semiconductor by replacing some original atoms with an atom of an element which by its nature has one more electron (and also an extra proton), the extra electron lives as a stranger when it is together with all the other atoms, but this time it is not compensated by a hole. Thus, I doped my semiconductor with an extra free electron, as it is generally done on silicon, as I already mentioned. Conversely, if I add an atom that by its nature has one less electron, the energy band that contains that atom has an electron hole, which is not created by an electron that has jumped into the next band, but which attracts electrons from the surrounding atoms that tend to fill it in: in short, we have the effect of having increased the number of holes.

In conclusion, this chapter has certainly not rigorously explained to you how diodes are created, nevertheless I hope you got the feeling of their bond to quantum mechanics. In addition, it should have made you aware of how diodes have revolutionized electronics. With the obvious deduction that, without quantum mechanics, we would never have gotten here.

7.5 – Quantum Entanglement and Friends

The interesting feature of "quantum entanglement" doesn't lie in the word "entanglement" but in the word "quantum".
Indeed, entanglement only means a strong correlation between two (or more) subjects, such as the opposite faces of a game dice, the sum of which is always 7. Or a pair of gloves, where if one is right the other is left, and so on.
Let's take the gloves case, for example: if take a pair, put one in a box and one in another, give the two boxes to two friends who leave for two different destinations and, when they arrive at their destination they open the box, I think you are not surprised if I say that each of them instantly knows which glove the other friend will see.
But if I have entangled quantum gloves - actually pairs of photons or electrons coming from a single source - each of them, at the start, is both right and left (remember the double slit experiment, where the electron passed from both A and B ??). So, once we arrive at our

destination, if you open the box this will affect the state of your glove (exactly like with the double slit experiment, when we measure which slits each particle goes through), forcing it to become either right or left. But above all, this will also affect the state of my glove, which in turn is no longer the superposition of two possibilities states, but only one, AS IF you had also performed the measurement at my place! Now I hope you have not missed the fact that this influence on the state of the other glove is instantly exerted, regardless of the distance between the two of us.

Perfect, we have found a way to perform actions at a distance at infinite speed.

Well, true and false: it is true that we have instantly influenced what happens at the other site, and we also know what will happen. Unfortunately, recalling what we learned on quantum mechanics, we are the only ones to realize this, therefore the possibilities to exploit this feature are very limited.

Let's see why I say that we are the only ones to notice it. Let's just assume you, who are at the other site far from me, open the box with a billionth of a second delay (even less) and see the glove in only one of the two states: this is normal, in fact it's also happening to me when I open my box. Indeed, quantum mechanics tells us that you or I will see only one of the possible states, once you or I perform the measurement. So you have no way of knowing if your measurement is the result of your or my action or else.

This means that there has been no transmission of the information that I have done a certain thing, even if it is really true that our action in some way has influenced reality from a distance.

But some attempts to exploit this feature really exist, let's see one.

7.6 – Quantum Cryptography

None of us is an encryption expert, but I think it is clear to everyone that in order to exchange encrypted messages, two subjects must have agreed upon a code, which is called a "key". They use this key to alter the message at the start and then decrypt it upon arrival. Well, quantum entanglement allows you to decide this code whenever it pleases you, and exchange it in a secret and secure way, that is, without the risk of someone intercepting it.

Our ingredients will be quantum mechanics superposition of states (the glove is both right and left) and Heisenberg's uncertainty principle. Thus if I know the orientation (right or left) of the glove, I cannot say whether it is made of wool or of cotton: actually, it is both wool and cotton. Conversely, if I know it is wool (or cotton), the glove is both right and left. Okay, when I open the box I can only see one of the two features and each time I can decide which one I want to see.

So let's have a source that sends us many pairs of entangled quantum gloves, one at a time. Each time the gloves reach us, each of us opens the box and independently chooses each time whether to measure the fabric or the orientation. We have two scenarios: if we are measuring the same property, the two results will be correlated with each other (but so far we have no way to notice that), otherwise they will be a bit random. As I said, we don't know which scenario yet. Then each of us writes on a piece of paper what kind of measurement we made and what result we obtained. After we make a certain number of measurements, we stop and we inform each other, over a normal telephone line, about the type of measurements (fabric or orientation) we have made and in what order, without revealing the result. Now we both know which measurements were correlated and which weren't, we discard the uncorrelated ones and we know exactly what result each other measured and in what order (without having said it, but only by exploiting entanglement). For example, if it turns out that we have only two correlated measurements (extremely simple case), which were respectively fabric and orientation, each of us knows which fabric our friend has seen with the first selected measurement and which orientation with the second.

We thus created a secret code (the sequence of two data), we are the only ones to know it, and we can now use it to encrypt and decrypt messages. And we can repeat it again, as often as we want.

I know what you think: what if a third person got in the way and eavesdropped on our gloves exchange process? Well that person would have performed the measurement and would have influenced the two gloves, by destroying the entanglement between the measurements later performed by us. So, by exchanging ALSO the results of a subset (not all of them) of the potentially entangled measures, we will realize that someone has tried to intercept our secret communication when we see that some of them are not entangled at all (I remind you that our data are no longer entangled once the eavesdropper has performed the measurement).

I hope that's enough.

Chapter 8. From Waves to Strings

String theory is one of the most promising theories that would allow to reconcile quantum mechanics with general relativity, and even to do much more.
This yet to be proven feature is important because, as I will tell you shortly, physicists have big problems in putting together these two theories which are the best available so far, but you cannot make them coexist.
In short, the "Standard Model" distinguishes all the elementary particles from one another upon the value of some properties such as mass, electric charge, spin and a few others. In string theory those properties turn out to be the manifestation of particular vibrations of a unique entity that we call "String".
But don't be fooled, I am not talking about the vibrations of the waves that quantum mechanics associates with the state of a physical system. In that case the system already has defined values for those properties (mass, electric charge, etc.) and the wave is used to describe the probability that other properties assume certain values that eventually I could measure, such as speed, position, spin orientation or others.
This string would constitute the one and only basic ingredient of the entire universe, as opposed to what we now believe to be the various fundamental ingredients, electrons, photons, quarks. Those ingredients, according to string theory are only the manifestation of a particular vibration of the String.
This theory, as I mentioned earlier, would have the advantage of describing in a unified way all the particles and all the fundamental forces of the universe discovered up to now. I use the term "would" because to date there is no experimental evidence that confirms its validity (but not even that disproves it) and scientist haven't even been capable to hypothesize a

single experiment that can try to probe the validity of the theory.
So let's go back to the theory alone, for now.
First question, but how big are these strings?
Infinitely smaller than an atomic nucleus, about 10 to minus 34 meters, therefore hundreds of billions of billions (yes twice !!!!) smaller than the nucleus, and, needless to say, since they are the fundamental ingredients of the universe, they themselves are not are made of anything smaller. I mean, contrary to an ordinary string that can be broken down into small pieces, up to atoms, indeed to protons and neutrons and electrons, or rather to quarks and electrons, all the way down to strings if you wish, strings have no ingredients that make them up. I realize that it is difficult to visualize because on the one hand we have said that these are not a dimensionless point-like object, they have a length, on the other we say that they are not made of other smaller things.

To make this concept even more complicate to visualize I recall that these strings even vibrate and it is their vibrations that we perceive as particles.
How is that?
Well let me make a simple case. If we remember Einstein's formula $E = mc^2$, and if we remember Planck's formula $E = hf$ which says that each vibration corresponds to a certain value of energy, we can say that to each frequency f of vibration of these strings corresponds a certain energy, which however is equivalent to a mass and in fact we perceive it as mass. In the end we perceive the frequency of vibration of these strings as mass.
Okay, but particles don't just have mass. Well here the theory becomes more difficult to trivialize but we can say that a wave is not characterized only by its frequency. Let's make an analogy with sound. The musical note "a" is characterized by a precise frequency, but the sound of "a" is different if it is produced by a piano or a guitar, or even a flute. Likewise, there are other properties of the vibration of this string each of which we perceive as one of the other intrinsic properties of the particles (electric charge, etc.).

Ok but why make life so complicated?
But because, as I said, with this trick perhaps we can also describe the force of gravity in a quantum way, which today is the only force that we manage to describe "only" with general relativity.
Instead we already describe the other three forces in a quantum way, the electric force, which was originally described in a classical way with Maxwell's equations but then, as we saw for example with the photoelectric effect, we were forced and managed to *"make the quantum leap"*, and the two nuclear forces.

But why can't we do it with gravity? The problem is that the formulas we use to describe the force of gravity depend on the distance between objects, but when we talk about quantum objects the concept of distance becomes at least evanescent. In fact, we have seen that the position of a quantum object is not defined, therefore we must find a way to overcome this strong dependence on the property of a precise distance. With the electric force we have succeeded, changing the paradigm, imagining that it is due to the exchange of one or more photons and then with this photon exchange mechanism managing to reconstruct the classical expressions that describe the force. But General Relativity is deeply dependent on the properties of space itself, as it models the gravitational force as the effect of the distortion of space time due to a certain mass in a certain precise position. Therefore, if according to quantum mechanics the mass now no longer has a precise position, I cannot predict how my space-time is deformed and therefore what force of gravity there will be. We should therefore find a way to model gravity as the exchange of particles, to which we have already assigned the name of gravitons, in the same way as we managed for the other three forces.
In the case of the electric force we have succeeded thanks to the fact that initially we had an experimental phenomenology (for example precisely the photoelectric phenomenon) that showed us the way. But with gravity, given its low intensity, it is not possible, as I said before, to even hypothesize what kind

of experiment could show us the behavior of this force in cases (small distances) in which Einstein's theory certainly will fail.
Why would it fail?
Working just with theory, no attempt managed to eliminate the presence of some distance in some denominator, starting from Einstein's equations.
And what does that mean?
It means that when the distances become zero, which is possible if my ingredients are all point-like as in the standard model, all the quantities involved become infinite and therefore Einstein's equations do not allow us to predict how gravity behaves over small distances.

So what about Strings?
Well we observe a bizarre coincidence, if it is a coincidence, let me explain. These strings were initially hypothesized only to describe the strong nuclear forces, but then going to dissect the mathematical properties of the equations that described them, we got for free some modes of vibration that seem precisely suited to being able to be interpreted as the gravitons, that is the envisaged sibling of the photon, and that would help us so much to reproduce in a quantum way the predictions of general relativity.
And that makes sense, because the problem of infinities at zero (or small) distances is overcome precisely because the string has its own size and therefore we would never go below a certain distance, just enough to avoid the problem. In other words, once the strings have been introduced, their "extended" nature leads to the definition of a minimum scale with which we can / must consider the whole universe. This necessarily leads to changes in the equations of general relativity with the result of overcoming the problem of infinite.
And now let's move to a more intriguing byproduct.
How do we manage to describe the behavior of forces with these strings?
The answer is with the twisting of the strings, which, as we will see shortly, eventually leads us to believe that the universe is composed of more than the 3 spatial dimensions and a temporal dimension we know today, but let's proceed in order.

Let's start with this "twist".
Ironically, the idea, for all forces, comes precisely from general relativity.
I recall once again that the force of gravity is described by Einstein's equations as the effect of a distortion of space-time. I say "I RECALL" because with the recent experimental confirmation of the existence of gravitational waves I am sure that all of you have attended at least one seminar on this topic, where they have surely told you that for example the Earth revolves around the Sun because the Sun with the its great mass deforms the space around it to the point that what without the sun would have been a straight line has become a circle or rather an ellipse. And then the Earth that is convinced to be traveling along this straight line actually finds itself traveling along an ellipse and we interpret this as if there were a force acting on the Earth directed towards the Sun.

Drawing on this approach, we can claim the following:
Let's pretend for a moment that we are in one dimension only, better yet we believe that we are in only one dimension. Yes, because in this example our universe is not just a line, as it would be in the case of a single dimension, but it is a tube where the circumference of this tube is so small that we do not even realize it exists. Now if the distortion of space transforms this tube into something like a thread of a screw and this helical can turn on itself, now if I have two of them interlaced to each other, I can have them to move towards each other or to pull away depending on how they turn. But, as I said, I don't notice the presence of this thread because the circumference is too small. I only see my two supposed one-dimensional strings approaching or moving away. It is precisely in this way that it was possible to describe the electric force.
But here you can say, sorry, the example you gave is possible because you assumed only one dimension for the string and you used one of the other two to transform the string into a tube. But we are already in three dimensions, in which of the three would the string be twisted to generate the electrical force in the real world?

The answer is in a fourth spatial dimension, which in fact we do not see (and we cannot even imagine) because it is closed on itself, like the surface of a cylinder but with a circumference that is too small for us to be able to perceive it. In the end, with this trick, the theory needs 10 dimensions to our space-time, instead of the now canonical 4, but the additional six are all of this "strange" type, in jargon they are called compact.

Well, I think I might have created a strong headache to you and I stop here because the topic is still in great turmoil, starting from the theory itself, because we still have a significant amount of "dots on the I's" to put, and also because we still need to design at least one first experiment that allows us to verify if this theory works or not.

Chapter 9. True or False? Many Universe Theory. The Law of Attraction. God's Equation.

I am forced to tackle three topics to help the reader distinguish what is likely to be a sound concept, the Multiverse or in any case the concept of Parallel Universes, from the other two that have no scientific grounds, the so-called " Law of Attraction" and the alleged "God's Equation"

Actually, had I managed to write this book clearly, this appendix would be redundant, because you, the reader, would now be able to make this distinction by yourself.

But since it is very unlikely that I was able to clearly express everything I wanted to convey, I must dwell on these issues to try to provide you with the tools to avoid being scammed or swindled.

Let's focus on my expression "*...have no scientific grounds*".

I'm not saying the topic is unlikely or even false. I'm just saying it doesn't have scientific support. I guess I'm still not clear on this.

Allow me to rephrase the concept with an example: string theory too (see chapter 8) cannot currently be considered a theory with sufficient scientific support. I remind you that to date we are not yet able to even imagine how to design an experiment that can be performed to test its validity.

So what? Why this double standard that made me talk about string theory, while some of these topics, it is already clear, will turn out to be no good?

That's a legitimate question, and I may answer: string theory at least has some formulas behind it, which was developed starting from the formulas of theories that have scientific grounds. But I don't like this answer and I hope neither do you.

Come on, stop messing around, you will now complain irritated by this apparent schizophrenia of mine.

What I really want to stress is that it even if we manage to express a theory or a model by means of a mathematical formula that doesn't give it scientific dignity; but at least the presence of a formula takes us a step forward towards the possibility of designing some experiment that can evaluate its validity. So when we do not even have a formula we have the CERTAINTY that it will never be possible to verify the scientific validity of the theory and at that point everything can turn out to be valid, no one will ever be able to say that it is wrong, but this does not mean that it is true.

In this chapter I want to outline which of those three themes at least start from scientific grounds and which doesn't, although many writers try to convince their readers that there is a link with quantum mechanics and /or with the theory of relativity or even with cosmology.

9.1 Multiverse or Many Universe Theory

Ok, let's start with the first one, which is also the only one to pass my filter: Multiverse or the Parallel Universes.

Multiverse, or Parallel Universes is nothing more than an attempt, certainly fascinating, to provide quantum mechanics with an alternative interpretation to the Copenhagen one, which I have extensively talked about in chapter 5.

Personally I don't even consider it that much alternative, but let's see what it says and then we'll make some comments.

Basically it is argued that when we perform a measurement on any physical system about the value of a yet undefined property, the very act of measurement involves the "cloning" of the universe into as many universes as the number of possible outcomes of this measurement. For example, the "physical system" may be the electron that started from our "rifle" and is about to reach our photographic plate after crossing the two slits obstacle, the "property" we want to measure is the slit it passed through, so our measurement will create two universes. Both the observer and the observed system and the whole universe are replicated and differ only because in each of these replicas the property we are observing assumes exactly one of those possible outcomes, in one universe the electron has passed through slit A while in the other one it passed through slit B (I just chose A and B to name the two slits).

A little while ago I said that "*I don't even consider it that much alternative*" because this approach does not alter any of the formulas of quantum mechanics, but merely provides an interpretation that perhaps for some people is even easier to "digest" than the Copenhagen one, where the universe remains only one but what changes is the property of the system, which is forced by the measurement to adopt one and only one of those possible values, a phenomenon that in jargon is defined as "collapse of the wave function".

I will not go further into all the features of the Multiverse, which as I said is just a different way of seeing things. I will focus only on one aspect where, in my humble opinion, the Multiverse is more elegant than the Copenhagen interpretation: quantum entanglement and the consequent apparent action at a distance.

Do you remember what we are talking about? If we have our quantum gloves we can measure only the orientation (i.e. if it is a right or left glove) or only the fabric (if it is cotton or wool) but not both together, and this comes from Heisemberg's uncertainty principle. If Bob measures the orientation (or the fabric), immediately also Alice's glove becomes a glove with a defined orientation (or a defined fabric). This, as I explained to you, seems to violate the principle of relativity which prevents immediate actions at any distance, since according to relativity we need to wait at least to let light travel from Bob to Alice before having any effect on Alice's glove. I also remind you that I told you that this violation is only apparent since no information travels because Alice cannot know if the result of her observation is due to what she did or to what Bob did, but let's forget about this last comment for a moment.

Using the Multiverse approach, the violation is not even an issue, given that what happens is that all possible universes are created (4 universes in this case), in each of these universes the two gloves have always had one of the two property already well defined and consistent between the two; the action of the measurement has therefore not altered any state of the gloves but rather has seen the Alice and Bob enter one of these universes, indeed 4 instances of Alice and Bob pairs (left/right, right/left, wool/wool, cotton/cotton), one in each of them, but none of the observer's "clones" will have the opportunity to come into contact with the others, each will continue their life in their new universe and so on.

9.2 The Law of Attraction

Surfing the Web about this "Law of Attraction", I can try to wrap-up its principles with these five statements:

1. Quantum Mechanics describes both matter and forces in terms of waves;
2. Each physical system, including yourself, can/must be associated with a wave that pervades all of space and time;
3. The wave describes at each point of space / time the probability that a certain property takes on a certain value, including for example the position of the object itself.
4. Only when a measurement of the value of this property is performed, the object will assume one of the values allowed by the wave (the collapse of the wave function).
5. Hence, since the measurement is an act of our conscience, we can use this option to make the measure converge towards values favorable to us thus turning reality to our own advantage.

From here on, with a leap that is not very clear to me, they go on to say that if we keep a certain positive attitude we have the consciousness that is better tuned to the waves of reality and is therefore able to influence the outcome of the measurement to the point that "... we can better manipulate our physical reality"

Let's comment together.

The first 4 points are all correct!!

Point 5, on the other hand, is already a symptom of a certain confusion as there is no need for consciousness to perform a measurement; please allow me to use the following example.

To quote Einstein himself, in one of his attempts to challenge the Copenhagen interpretation, it is said that Einstein said to Bohr something like "I like to think that the Moon is there even if I don't look at it".

What did Einstein mean with this sentence?

He was convinced, as we all are in our day by day life, that all the properties of a system are defined, so he was strongly disturbed by the fact that according to quantum mechanics the reality we live in is created by the act of measurement. So he fooled around pretending to fall into the same misunderstanding we see in point 5, that the measurement is carried out by man and therefore paradoxically if none of us looked at the Moon it would not stay where it is.

In quantum mechanics, the concept of measurement is the interaction of a system (both quantum and classical) with a macroscopic system. Carrying on with Einstein's argument, the Moon actually interacts with the multitude of photons arriving from the sun or any other star in the universe. Each of these photons "observe" a certain value of the position of the Moon in the sky, but since according to quantum mechanics the state of the Moon is such that there is a value of its "position" property that has greater probability than all the others, the Moon turns out to be right where it is, even if no human being was watching it. Likewise, if in place of the Moon I had a single electron, all those photons would determine its position. The difference with the Moon example is that with the electron, measurements in successive instants of the position would not allow me to estimate a uniform circular motion along a certain orbit, which instead happens with the measurements performed on the Moon object. For the electron, the measurements would give random positions, albeit always and only in that circumference that we call orbit[1]. In essence, therefore, it is certainly not quantum mechanics that tells us that "... we can better manipulate our physical reality". This does not mean that it is not true that having a positive attitude ends up facilitating things and all the other amenities of this alleged law of attraction. But we must not fall into the error of believing that it is quantum mechanics that gives us further confirmation (or denial per se)!!!!

Let me take this opportunity to also address the issue of consciousness and quantum mechanics. It is in fact true that deterministic science has always created a moral problem for scientists. In fact, according to classical science, everything is or would be predictable thanks to the application of the laws of Mother Nature, as long as we know the initial conditions. So even our individual choices and / or decisions are actually already predetermined, only we do not know, because we do not know all the laws of Mother Nature and even less do we know precisely all the aspects of the initial conditions. For example, let's go back to the usual dice: the outcome of the throw is actually predetermined! It is only our ignorance that does not allow us to predict it accurately and then we resort to statistics. According to classical science this applies not only to the dice but also to our free will which turns out not to be free at all, it "feels" free just because we are not able to predict it. The question cannot but disturb us, and it is therefore obvious that with the arrival of quantum mechanics, where the outcome of the quantum-dice roll cannot be predetermined, a loophole has opened up to try to free our free will (forgive the pun) from the constraints of classical determinism. World-renowned physicists like Roger Penrose have also embarked on this adventure. To date, they too have only elaborated

---

[1] Please be advised, what I have just said is not an interpretation, but is precisely the result of the rigorous application of the equations of quantum mechanics. I remind you that in paragraph 6.5 I wrote that if we apply the equations of quantum mechanics for example to predict what happens to a bowling ball, rather than to a single electron the results would be exactly the same as the equations of classical physics, therefore we keep using classical physics to predict the behavior of a bowling ball because its simpler.

theories, which are fascinating, but which still lack experimental proof. So I am not saying that they are not true, but I am saying that at the moment, these too are only theories.

9.3 God's Equation & "Partners"

In this paragraph we examine very briefly a collection of themes that are all united by the term "God" or by the term "everything" and therefore are often confused with each other:
God's Particle
God's Equation (books)
God's Equation (math)
The Theory of Everything

The God's Particle

This one is simple because it's just a transcription error, which however shows how strong is the desire to use these high-sounding terms always finds a way to assert itself.
The so-called "God Particle" is none other than the famous (famous?) Higgs boson, which made the headlines in 2012 when it was first detected thanks to two CERN experiments.
Why this name?
Absolutely no reason, except that its existence had been predicted in the sixties by a group of theoretical physicists, including the Scottish Peter Higgs from whom it takes its name, and since then we have never been able to be observe it, up until 2012.
But it seems that it was precisely this difficulty in finding it that caused one of the legends of particle physics, prof. Leon Lederman, whom I also had the privilege of knowing personally when I was working at Fermilab in Chicago, to use the expression "...this goddamned particle...". After that the word of mouth turned it into " God particle" and the rest is history.

The God's Particle (books)

There are at least a couple of books that use a similar title, whose noble purpose is to illustrate to the reader the hypotheses that are brewing in physics in an attempt to unify quantum mechanics and Einstein's theory of general relativity. You yourselves have had a brief account of it in this book in chapter 8 devoted to string theory.

What I therefore want to reiterate is not to fall into the misunderstanding of finding a text that explains how this phantom unique equation of physics is made or how it was born, because it does not exist, and above all, even when we should be able to formulate it for the first time, it will never be the "final solution", forgive me for this expression which unfortunately also recalls horrendous historical facts.

Remember that physics, actually all the sciences, will never get to the "right" equation, but only to equations that are increasingly powerful and reliable. We must always remember what Plato said with his myth of the cave: we are in a cave facing inwards, and we can only see the shadows of what happens outside.

The God's Particle (math)

Here the topic becomes much more interesting but we leave the field of science and enter the field of mathematics, and it is precisely for this reason that it becomes interesting.

WARNING: I remind you that I am (was) a physicist, who is about to venture into a fascinating field of mathematics, which is always out of physicist's reach.

Why do I say *it gets interesting because it's math and not physics*?

Because mathematics, unlike the natural sciences, is a truly exact discipline. And I used the term "discipline" and not "science" precisely to affirm its superiority.

If in mathematics a relationship is true, then it means that it is really true !!!

The so-called "Equation of God" that we will now deal with, is therefore an absolute TRUE relationship, exactly like saying that 2 + 2 is 4, and it is not linked to any natural phenomenon. But then what is special about this equation as opposed to 2 + 2 = 4?

As we will see shortly, it expresses a relationship that appears to exist between three very complicated numbers, numbers that are called "irrational", indeed one even "imaginary", and which are the number PI π, Euler's number e, and the imaginary number i, with the simplest of numbers, the natural number 1, and the number 0; in formulas

$e^{i\pi}+1=0$

I'm pretty sure it doesn't impress you much, so far, and so I'll try to make you appreciate some aspects of all numbers that you may never have noticed and which I hope will allow you to feel empathy towards those who have been so amazed by this relationship that they call it the Equation of God despite its real name is Euler's Identity.

What I will try to accomplish is to make you appreciate that numbers are apparently just a language. A language artfully created by mankind to represent reality. However, unlike all other languages, the language of numbers has very few rules imposed a priori, all the rest of its properties are consequences of these rules. In addition, as I said before, these properties, which have been gradually discovered, are absolutely true, they can never be denied. So the discoveries of many of these absolute truths lead us to imagine that "mathematics brings man closer to God", I don't remember where I read this sentence but I really like it, precisely because it leads us to discover absolute truths. And among all, the beauty of this equation stems from the connection of two very simple but very important numbers, which are the number 1 (the neutral element of the multiplication operation, in fact if I multiply a number by 1 I get the same number as before) and the number 0 (the neutral element of the addition) with three very complicated numbers (soon we will see what I mean) and which are also the protagonists of all the most important equations of science. Having discovered that there is a link between these two groups of numbers cannot fail to trigger the suspicion that this relationship expresses, with the language of mathematics, a fundamental absolute truth, even if we do not know which one.

So for those who haven't already got bored, let's dive into the wonderful world of numbers.

The first set of numbers that come to mind, and which are the ones with which humanity started from the beginning, are the numbers that today are defined as "Natural". Number 1,2,3, etc.

These numbers represent the reality we observe every day. We see one object, two objects, etc.

Natural numbers turn out to have unexpected properties. I point out that one of these is the fact that these numbers exist absolutely. I do not need to see two objects to say that for example the number 2 exists. On the other hand, this object allows me to represent an infinity of real situations, those where I have two objects, or even where I have counted two successive events and so on, it is very powerful.

I can define an operation which I call "sum" that allows me to create any other natural number by adding two together and then I also discover that there are also properties of this operation that are called commutative property and associative property; do not worry if you do not remember them, we just need to know that some properties exist and that these properties have not been decided at the table, but they turn out to be a consequence of the very simple procedure with which we created these entities that we call "numbers" or rather "natural numbers ".

That being said, I hope you agree with me that you can create any natural number by adding two other numbers together, with the exception of the number 1. The number 1 for now cannot be obtained by adding other natural numbers. Then we could also introduce another object, the zero, and find that it has at least the same properties of the numbers that preceded it and we have solved the problem of the number 1, but in reality we have only moved the problem to the number zero.

So we introduce yet another object, we call it "-1" and we say that it has the property that if added to the number 1 it gives the number "zero". Once this is done we discover that this object "-1" has the same properties of the previous numbers provide that I also introduce the numbers -2, -3, etc. I call the set of these new numbers and the Natural numbers "Integers" and this time I find that I can really get any "Integer" number as the sum of two other Integers and that the number 0 is the neutral element with respect to the sum operation, while the number 1 is the neutral element of the multiplication operation which however is only a combination operation of sum operations.

Okay, but what do these "Integer Numbers" represent? Have you ever seen -5 horses? I don't think so!

As a workaround you could say, for example, that with the temperature we also have -5 degrees, but if I move the zero position a little lower I go back to the Natural numbers. However, this attempt gave us a good suggestion because in geometry I can say that these numbers represent the quantity and the direction in which I move. If I take a segment and say that its length is 1, then I can say that if I lean to the right I move by 1, if I lean to the left by -1 and so on. We have therefore found at least one real situation that can be represented by these integers. I would add that physics will provide us with a second case, when with electrical phenomena we are forced to introduce a concept called Electric Charge for which we are forced to hypothesize that there are two, and by resorting to the signs of numbers it is possible to give a quantitative representation of this aspect and all its consequences.

Now let's get back to geometry. I hope you agree that it should ALWAYS seem possible to trace a segment of a certain length and then trace a smaller one that is contained in the previous one a natural number (in everyday talk we say "whole" instead of "natural") of times. If we now say that the longest segment is the unit length, we could ask ourselves if there is a number that can represent the length of the smaller one or two small ones, etc. Well we would realize that this number exists, it is none other than the fraction $1/n$, where 1 is the natural number 1 and "n" and the natural number n of times that that small segment stays in the longer one.

Attention, as I said, we REALIZE that these objects we have defined "*1 / n*", and then also "*2 / n*" and then again "*(-1) / n*" or "*(-2) / n*" or "*1 / (- n)*" etc. have exactly the same properties as numbers and therefore they too are a type of number. We call them Rational Numbers all these numbers. Integers and Natural turn out to be just a special subset of this Rational Numbers.

The Pythagoreans hoped that it was always possible to draw two random segments and find out that the relationship between their lengths is a Rational Number. If this were the case it would mean that with rational numbers we can express the relationships that exist between any pair of segments and by extension between any object of our universe.

But they themselves soon realized that this is not quite the case. It is necessary to further expand the family of numbers because already in geometry there are cases in which, taking a couple of segments, Rational Numbers won't succeed.

I will mention two of these cases, one is certainly well known to you, the other perhaps just as well, but you have to brush-up a few more scholastic aspects.

The first one that I mention concerns the relationship between a segment and one as long as the circumference of which the first segment is the radius or the diameter. The relationship between these two segments contains a new number, which cannot be expressed as a ratio between two integers. It is the PI number π.

The second case is the square root number of 2, which you can construct by drawing a square where it turns out that the relationship between the side and the diagonal cannot be expressed with a rational number, it is in fact the irrational number that we call the "root of two".

It turns out that the "π" and "root-of-two" objects once again have the same properties as all other numbers introduced so far. In short, it is necessary to introduce these numbers which cannot be expressed as ratios between integers, and which are called irrational numbers.

A peculiarity of these irrational numbers is that not only can we not express them as a result of a division between two integers, but we cannot even write them precisely, because

after the comma they provide an infinite sequence of digits which, however, will never present any regularity.
What do I mean by that?
For example, I can write the rational number 1/5 using the expression 0,2 and only two digits were enough. Then if I get more fancy I may tackle rational number 1/3, I can write it 0,33333333.... with an infinite sequence of 3, but there is only the digit 3, then there are other rational numbers that maybe come out as 0,acbdacbdacbd ... where we have an infinite sequence but always and only of the "period" (this is called) "acbd ". For irrational numbers, on the other hand, not only we will have an infinite sequence but we will never know which digit is to the right of the last one that we have somehow managed to calculate, we have to redo an even more precise and complicated calculation.
With these irrational numbers we should be done. Reality can be represented by the set of rational and irrational numbers which, not surprisingly, are called the set of Real Numbers.
Needless to say, we didn't stop. At a certain point, as a game, mathematicians ask themselves if an object that they define as "square root of minus 1" is a number, and it turns out that this is the case. They assign it the symbol "i", and all the numbers that they are multiples of these "i" are called Imaginary Numbers and all numbers that are a mixture of Imaginary and Real are called Complex Numbers, which can be expressed as a + ib, where a and b are two Real Numbers.
These Imaginary Numbers, and even more so the Complex Numbers, seem to have absolutely nothing to do with reality, which instead seems to be represented only by Real Numbers. They seem just a game for mathematicians, but then quantum mechanics arrives and Complex Numbers turn out to be the basis of all the equations of quantum mechanics, suit yourself !!!
Having made this long introduction, I finish by presenting the number "*e*" called Euler's number which is defined as the result of this formula below:

$$e := \sum_{n=0}^{\infty} \frac{1}{n!}$$

Here that strange symbol called "sigma" just means that I make a sum for n ranging from 0 to infinity of what is written on the RIGHT of sigma; and that exclamation point to the right of n means that I take the result of multiplying and all natural numbers from 1 to n, in other words the formula says that

$$e := \frac{1}{0!} + \frac{1}{1!} + \frac{1}{2!} + \frac{1}{3!} + \frac{1}{4!} + \cdots,$$

where "0!" it is imposed that it is worth 1, while instead 1! is by definition worth 1, 2! is by definition worth 1 times 2 = 2, 3! 1 times 2 times 3 = 6, etc.

You will be wondering "but who the heck spent time to invent this stuff" and you are right! However, this number turns out to have an infinity of very useful properties in science, plus the following one

$$e^{i\pi} + 1 = 0$$

Don't you also think now that there must be something big behind this incredible relationship?

9.4 The Theory of Everything

We have come to the fourth and last of the topics I mentioned. I would like to point out immediately that various aspects lurk under this expression as well.
In fact, the variegated provisional results produced by the efforts with which the scientific community is working to try to unify Quantum Mechanics with General Relativity are referred to as the Theory of Everything, and one of these results is precisely the case of String Theory of chapter 8.
But it is also the title of a biographical film on the physicist Stephen Hawking and it is also a book, informative and non-biographical, written by Hawking himself, and it is precisely about one of his results that I want to briefly talk to you, because this theory of everything in reality does not exist, but

the results obtained by this physicist are bizarre and deserve an in-depth study.
We will BRIEFLY talk about the Hawking Black Holes Radiation.
I'm pretty sure you folks are all familiar black holes or that you have heard that they are called "black" because nothing can escape them, not even light, and that this property derives from Einstein's equation of general relativity which predicts that large masses curve space time etc. etc.
Well I am almost certain, on the other hand, they almost never told you, and neither did I in this book, but I plan to do it in one of the next ones in this series, that in 1928 one of the greatest physicists of all times, the English physicist P.A.M Dirac elaborates an evolution of quantum mechanics by incorporating in it also Einstein's Special Relativity (not the General, the Restricted, the one who among many things also says that $E=mc^2$) thus producing the vision still used today of quantum mechanics which takes the name of Quantum Field Theory.
Quantum Field Theory is not just a theory, as I said it has found millions of experimental confirmations, among which I mention the existence of **antimatter**, discovered experimentally in 1932 and I mention the fact that according to this theory it is perfectly normal that even in total vacuum, phenomena of creation of pairs of matter and antimatter particles occur. The life-time of these pairs is inversely proportional to the energy associated with their existence, so in more or less short times the particles disappear and the vacuum becomes empty again. Perhaps they have shown you graphic simulations of this where the void is not stationary but it is a bubbling of these particles who are born and die all the time. I point out that the particles of each pair are entangled, like our pair of gloves, and this has probably led many to claim that Dirac equation is the **equation of love**, but we will talk about this in the next book, let's go back to Hawking and **black holes**.
So what does our prankster (Hawking) do in the early 70's? He says *"suppose that near the event horizon of a black hole (you go and look for this on wikipedia) one of these pairs is created,*

but that instead of annihilating immediately, one of the two particles is immediately attracted to the hole black and end up in it! ". On the other hand, we are so close to that point of no return that some particle will end up in it every now and then. The other particle no longer having its companion will remain free to move in space like any other particle. This phenomenon is called Hawking radiation and at the moment it is only theory

I can try to calculate, Hawking wonders, the combined effect of these two events, one purely Quantum (the pair creation form void), the other purely Relativistic (the black hole attraction). He dives into the challenge and eventually comes to terms with these two theories and concludes, now I trivialize, that everything happens AS IF it were the black hole itself that emits the particle that escapes, and this should already surprise you because it would mean that you admit that something can come out black holes, albeit due to quantum mechanics; but the surprising thing is that the formula that comes out is practically identical to that of the black body radiation that I told you about in chapter 3 !!

Many scientists think It cannot be a coincidence, and therefore believe this is the way to follow to try to find this famous equation that puts together the two starting points (Quantum Mechanics and General Relativity) and not only for this particular case. PLEASE BE AWARE, even with Hawking Radiation, we are only talking about theory. No one has ever seen experimentally for now this Hawking radiation coming out of some black hole, even if recently some laboratory experiments (laboratory black holes? Yes, laboratory) have managed to reproduce conditions that we could say conceptually equivalent to those of a black hole and to observe that something actually comes out of it.

https://www.theverge.com/2016/8/17/12514796/black-hole-radiation-stephen-hawking-proof

Conclusions

I hope that by now you know what this phantom quantum mechanics is.
We have, I hope, learned that it is the theory with which science can describe Nature's behavior at the "microscopic" scale, where classical physics instead failed miserably. I used quotation marks because in reality it is on a much smaller scale than simply "microscopic" that the need for this new theory emerges.
And I hope we won't make the mistake of believing that the theory works for just that scale. In reality, it also works at our macroscopic scale, only its equations magically become identical to those of classical physics, so it is superfluous to start from quantum mechanics when we deal with "normal" objects.
OK but what's so special about it?
Again, I hope it is now clear that the price we had to pay for this powerful theory is to abandon the deterministic view of the reality we live in. In fact, although in our everyday life everything appears to be described in a precise and punctual way, when we instead investigate the properties of the individual basic ingredients of our entire universe, their nature turns out to be intrinsically statistical. And when I say statistical, I do not say it in the classical sense: meaning that when we have equations that are too difficult to solve, we are forced to focus only on the average of their results because it is a simpler problem; or when I cannot measure all the initial values of all the parameters for which my classical equations would provide precise predictions, I am happy to look only at the average values.
"Intrinsically statistical" means that each basic ingredient of the whole universe has no definite property, at least until I perform the observation of these properties. And on top of that, I also had to give up the possibility that I can measure all of them at the same time.

In other words, if my classic dice, inside a closed box, has a precise face upwards and it is only my ignorance that makes me say that it can be in one of the six positions with a probability equal to one sixth, my quantum dice has no face upwards, of rather it has them all at the same time. But as soon as I open the box, it magically chooses one, at random, or - I should say - following a precise probabilistic law. And again, if my classic dice has a precise face facing right, my quantum dice, even after I observed which face is facing up, does not have a precise face facing right; if I measure which one faces right, then the upwards face becomes indeterminate again.

Quantum mechanics uses these absurd assumptions, and manages to describe the behavior of atoms so faithfully that we can predict the existence and properties of semiconductors and with them build the electronics that underlie all the objects that we use today.

However, history repeats itself and we can guess that, sooner or later, even this theory will fall into a crisis. It will be up to future scientists to find a new one, which will be even more powerful and precise than this one. However, at present, I have good reason to believe that it won't be me who will tell you about it.

Thanks for reading this Book!

We've reserved an **EXCLUSIVE BONUS GIFT** for You!

Download for **FREE** here : https://tinyurl.com/3a97bdpk

Or use this QR Code:

We also want to ask you for a minute of your time to
Leave a Review
about this book. It is very important for us to know what our readers think. This helps us to make better and better products.

Notes

Image Credits

Fig. 1.9 e 2.4 – (mod.): https://tinyurl.com/tm9u37r4 (CC 0)
Fig. 2.5: https://tinyurl.com/45ejvvdy (CC BY 4.0: https://creativecommons.org/licenses/by-sa/4.0/deed.en)
Fig: "Sphalerite" by Andreas Früh (Andel) - Wikimedia (CC BY-SA 3.0)
Fig. 5.1 – (mod.): Double-slit.svg" by !Original: NekoJaNekoJaVector: Johannes Kalliauer (CC BY-SA 4.0)
Fig. 5.2: Double-slit experiment results Tanamura 2.jpg" by user:Belsazar (CC BY-SA 3.0)
Fig. 5.7: Pixabay "Attribution not required"
Fig. 6.1: CC0
Fig. 7.2: "803 Vacuum Tube (size comparison)" by Nick Ames (CC BY-SA 2.0)
Cover: Pixabay (mod.) "Attribution not required"

Text Notes

1) https://www.youtube.com/watch?v=Iuv6hY6zsd0
2) https://www.youtube.com/watch?v=xH30jaDqi6c
3) https://www.youtube.com/watch?v=oae5fa-f0S0
4) https://www.youtube.com/watch?v=CYLalgh83C0
im1) https://assets.sutori.com/user-uploads/image/82542f22-c5a5-4bbb-a8ec-8c948d6d3466/90185edb2a6f218c91f9b54e2c80dac8.jpeg

About the Authors

Antonio Scalisi

Graduated in Physics at the University of Genoa and former researcher in Particle Physics at CERN in Geneva and FERMILAB in Chicago. Currently he is dedicated to scientific dissemination, both by lecturing and through his YouTube channel: https://www.youtube.com/user/asca6601

Karing Ship

The team of "Karing Ship" - the first Editorial Line of Azad Publishing Ltd. It is formed for the most part by professionals with twenty years of experience in Publishing and Media. Thanks to these skills, our Team is able to create (*if necessary, as in this case, collaborating with selected experts*) Quality Guides and Manuals to satisfy the desires and needs of the most demanding readers.

Other books by the Publisher

Food Truck Business Guide for Beginners

Grill IT – The Italian Way

Coloring & Activity Books for Aduts & for Kids

Self Help Journals

Craft, Hobbies & Home Journals

School Planners & Notebooks

(SAMPLE)

Halloween Coloring Book for Adults Vol .1 - Mandalas

(SAMPLE)

Camping Journal for Couples

(Sample)

Gardening Planner & Log Book

Discover All the Books at Amazon!

Search for the following Authors:

Karing Ship
Azartis Cafè
Caroline Skylar Moore
Gordon J Underwood
AlanaWhitchy
Nora W. Liberty

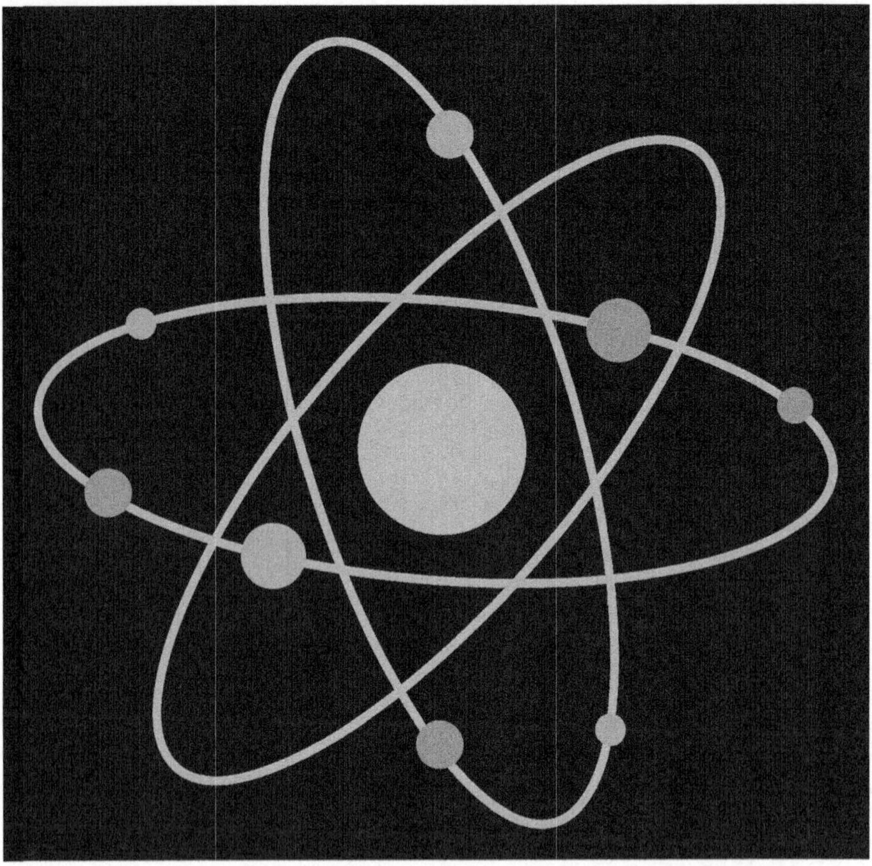